Dynamics of
Soil Organic Matter
in Tropical Ecosystems

Dynamics of
Soil Organic Matter
in Tropical Ecosystems

Edited by

David C. Coleman · J. Malcolm Oades · Goro Uehara

Published by:
NifTAL Project
Department of Agronomy and Soil Science
College of Tropical Agriculture and Human Resources
University of Hawaii

Library of Congress Cataloging-in-Publication Data

Dynamics of soil organic matter in tropical ecosystems: result of a conference held at the Maui Beach Hotel, Kahului, Maui, Hawaii, October 7-15, 1988, hosted by the NifTAL Project, University of Hawaii / edited by David C. Coleman, J. Malcolm Oades, Goro Uehara.
   p. cm.
Bibliography: p.
Includes index.
ISBN 0-8248-1251-4 (alk. paper) : $30.00
1. Humus--Tropics--Congresses. 2.Soil ecology--Tropics--Congresses. 3. Soil management--Tropics--Congresses. 4. Soil dynamics--Tropics--Congresses. I. Coleman, David C., 1938- II. Oades, J. M. III. Uehara, Goro. IV. University of Hawaii at Manoa. Niftal Project.
S592.8.D96 1989
631.4'17--dc20   89-12301
                                     CIP

This work supported in part by the National Science Foundation, Ecology Program, and the U.S. Agency for International Development, Bureau of Science and Technology, Office of Agriculture, through NifTAL (Nitrogen Fixation by Tropical Agricultural Legumes) and IBSNAT (International Benchmark Sites Network for Agrotechnology Transfer), College of Tropical Agriculture and Human Resources, University of Hawaii, in collaboration with the University of Georgia, Institute of Ecology.

Copy editing and design by *Cynthia L. Garver*, Calico Associates, Raleigh, North Carolina.
Final production, typesetting and layout by *Princess I. Ferguson and Susan Hiraoka*, NifTAL Project, Paia, Maui, Hawaii.
Chapter logos designed by *Rich Gabrielson;* art production by *Anna Gillies.*

Distributed by

University of Hawaii Press

Order Department

2840 Kolowalu Street

Honolulu, Hawaii 96822

Contents

Introduction: Soil Organic Matter: Tropical vs. Temperate

David Coleman, J. Malcolm Oades and Goro Uehara

Chapter 1. Constituents of Organic Matter in Temperate and Tropical Soils

Benny K. G. Theng, Kevin R. Tate, and Phillip Sollins
with Norhayati Moris, Nalini Nadkarni, and Robert L. Tate III

The first chapter illustrates the similarity of the chemistry of organic matter in temperate and tropical regions. Organic fractions are discussed on the basis of their stability or longevity in soils, and, it is stressed that the stability of organic matter in soils is largely controlled by the position of organic materials in the soil matrix and interactions with mineral constituents.

Chapter 2. Soil Organic Matter as a Source and a Sink of Plant Nutrients

John M. Duxbury, M. Scott Smith, and John W. Doran
with **Carl Jordan, Larry Szott, and Eric Vance**

The mechanisms involved in regulating the release and uptake of various plant nutrients are described. The major plant nutrients are contained in both above- and belowground biomass, but the cycling of nutrients is through the soil microbial biomass which thus plays a central role. Relationships between SOM and the supply of N, S, P, and other plant nutrients in temperate and tropical regions are discussed. Again, the role of mineral fractions in controlling biological and biochemical processes is stressed.

Chapter 3. Interactions of Soil Organic Matter and Variable-Charge Clays

J. Malcolm Oades, Gavin P. Gillman, and Goro Uehara
with Nguyen V. Hue, Meine van Noordwijk, G. Philip Robertson, and Koji Wada

Variable charge minerals are defined before descriptions of interactions of organic molecules with oxides in soils. These interactions control a range of properties such as retention of both anionic and cationic nutrients and also the strongly developed microstructure in variable charge soils. It is stressed that special techniques are necessary to study physicochemical properties of tropical soils, and that serious errors arise from application of techniques used in temperate regions. The importance of the maintenance of SOM in Oxisols and Ultisols are discussed.

Chapter 4. Biological Processes Regulating Organic Matter Dynamics in Tropical Soils

Jonathan M. Anderson and Patrick W. Flanagan
with Edward Caswell, David C. Coleman, Elvira Cuevas,
Diana W. Freckman, Julia Allen Jones, Patrick Lavelle, and
Peter Vitousek

The key variables regulating biological interactions in tropical SOM relate to resource quality (C/N, lignin/N, organic P + S, etc.) and edaphic conditions. Key biotic modifiers are plant roots and the dominant macro-faunal groups such as termites and earthworms. The latter can be quite active in certain areas which are seasonally wet, such as equatorial savannas. Anderson et al. emphasize that biological processes are qualitatively similar in temperate and tropical areas, but overriding factors such as soil texture and structure and presence of true geophagous fauna (especially certain termites), can markedly influence availability and distribution of nutrients in a given landscape.

Chapter 5. Organic Input Management in Tropical Agroecosystems

Pedro A. Sanchez, Cheryl A. Palm, Lawrence T. Szott, Elvira Cuevas, and Rattan Lal
with James H. Fownes, Paul Hendrix, Haruyoshi Ikawa, Scott Jones, Meine van Noordwijk, and Goro Uehara

Given the basic information presented in the first four chapters, how can one manage organic inputs to enable farmers in tropical regions to have flourishing sustainable agriculture? Sanchez et al. cite results of long-term experiments in parts of Amazonian and Andean Peru which give some initial information to address these major unknowns.

Chapter 6. Modeling Soil Organic Matter Dynamics in Tropical Soils

William J. Parton, Robert L. Sanford, Pedro A. Sanchez, and John W. B. Stewart
with Torben A. Bonde, Dac Crosley, Hans van Veen, and Russell Yost

The modeling group drew on information provided, in part, by others and addressed the question of universality of patterns and processes in temperate and tropical soil systems. They agreed that much more information is needed in addition to a general soil texture factor, which modifies such things as C/N or lignin/N values. The principal outcome seemed to be an increased desire and interest among researchers in tropical areas—both forested and agricultural—in developing more specific models for their particular regions. The utility of this approach for some of the Tropical Soil Biology and Fertility (TSBF) work was also noted.

Chapter 7. Methodologies for Assessing the Quantity and Quality of Soil Organic Matter

Frank J. Stevenson and Edward T. Elliott
with C. Vernon Cole, John Ingram, J. Malcolm Oades, Caroline Preston, and Philip J. Sollins

The ultimate questions of tropical SOM and its dynamics depend on suitable and readily available techniques for characterizing it, both in amounts and quality. This work demands that the investigator be aware of the physical nature of the soil and the ways in which physical fractions can then be subsequently characterized chemically. With our increasing concerns for sustainable ecosystem status and/or productivity, Stevenson et al. particularly emphasize fractions of pools of biological significance. The authors recommend using new techniques in situ, *unaltered examination of organic N, S, and P moieties using new technques such as ^{13}C-NMR, and pyrolysis-mass spectrometry. A great deal can be done, however, using the more readily available analytical tools of any soils laboratory to characterize the biologically meaningful pools of tropical SOM using a range of physical and chemical characterization techniques.*

Contributors

Dr. Jonathan M. Anderson
Department of Biological Sciences
University of Exeter
Exeter EX4 4PS United Kingdom

Dr. B. Ben Bohlool
NifTAL Project
University of Hawaii
1000 Holomua Avenue
Paia, Maui, HI   96779-9744 USA

Dr. Torben Bonde
Department of Water in Environment and Society
Linkoping University
S-58183 Linkoping Sweden

Dr. Edward Caswell
Department of Plant Pathology
St. John Building
University of Hawaii
Honolulu, HI   96822 USA

Dr. Vern Cole
Natural Resource Ecology Laboratory
Colorado State University
Fort Collins, CO   80523 USA

Dr. David C. Coleman
Department of Entomology
University of Georgia
Athens, GA   30602 USA

Dr. D. A. Crossley, Jr.
Department of Entomology
University of Georgia
Athens, GA   30602   USA

Dr. Elvira Cuevas
Centro de Ecologia
IVIC-Aptdo 21827
Caracas 1020-A, Venezuela

Dr. John Doran
USDA-ARS
116 Keim Hall, East Campus
University of Nebraska
Lincoln, NE   68583   USA

Dr. John M. Duxbury
Department of Agronomy
Bradford Hall
Cornell University
Ithaca, NY   14853   USA

Dr. Samir A. El-Swaify
Department of Agronomy and Soil Science
University of Hawaii
1910 East-West Road
Honolulu, HI   96822   USA

Dr. Edward T. Elliott
Natural Resource Ecology Laboratory
Colorado State University
Fort Collins, CO   80523   USA

Dr. Patrick Flanagan
College of Graduate Studies and Research
107 Warriner Hall
Central Michigan University
Mount Pleasant, MI   48858   USA

Dr. James Fownes
> Department of Agronomy and Soil Science
> University of Hawaii
> 1910 East-West Road
> Honolulu, HI 96822 USA

Dr. Diana W. Freckman
> Ecology Program, Room 215
> National Science Foundation
> Division of Biotic Systems and Research
> Washington, D.C. 20550

Ms. Cynthia L. Garver
> Calico Associates
> 809 Ravenwood Drive
> Raleigh, NC 27606 USA

Dr. Gavin P. Gillman
> Division of Soils
> CSIRO
> PMB Aitkenvale
> QLD 4814 Australia

Dr. Paul Hendrix
> Institute of Ecology
> University of Georgia
> Athens, GA 30602 USA

Dr. Nguyen V. Hue
> Department of Agronomy and Soil Science
> University of Hawaii
> 1910 East-West Road
> Honolulu, HI 96822 USA

Dr. Haruyoshi Ikawa
> Department of Agronomy and Soil Science
> University of Hawaii
> 1910 East-West Road
> Honolulu, HI 96822 USA

Dr. John Ingram
> Department of Biological Sciences
> University of Zimbabwe
> P.O. Box MP 167
> Harare, Zimbabwe

Dr. Julia A. Jones
> Geography and Environmental Studies
> University of California
> Santa Barbara, CA 93106

Mr. Scott Jones
> School of Biological Sciences
> University of Stirling
> FK9 4LA, Scotland

Dr. Carl F. Jordan
> Institute of Ecology
> University of Georgia
> Athens, GA 30602 USA

Dr. Rattan Lal
> Department of Agronomy
> Ohio State University
> 1735 Neil Avenue
> Columbus, OH 43210 USA

Dr. Patrick Lavelle
> Ecole Normale Supereure
> Laboratoire d'Ecologie
> 46 Rue d'Ulm
> 75230 Paris
> CEDEX 05, France

Dr. Norhayati Moris
> Rubber Research Institute of Malaysia
> P.O. Box 10150
> Postal Code 50908
> Kuala Lumpur, Malaysia

Dr. Nalini Nadkarni
Department of Biological Sciences
University of California
Santa Barbara, CA   93106 USA

Dr. Meine van Noordwijk
Institute for Soil Fertility
Oosterweg 92
Postbus 30003
9750 RA
Haren, The Netherlands

Dr. Malcolm Oades
Department of Soil Science
Waite Agricultural Research Institute
Glen Osmond, S.A. 5064, Australia

Dr. Cheryl P. Palm
Tropical Soils Research Program
NCSU Box 7619
Raleigh, NC   27695-7619 USA

Dr. William Parton
National Science Foundation
Division of Biotic Systems and Research
Washington, D.C.   20550 USA

Dr. Caroline M. Preston
Pacific Forestry Centre
506 W. Burnside Road
Victoria, BC   V8Z 1M5 Canada

Dr. G. Philip Robertson
W. K. Kellogg Biological Station
Department of Crop and Soil Sciences
Michigan State University
Hickory Corners, MI   49060-9516 USA

Dr. Ignacio H. Salcedo
    Rua Antonio P. Figueiredo, 140
    Boa Viagem
    51011 Recife-PE, Brazil

Dr. Pedro A. Sanchez
    Tropical Soils Research Program
    North Carolina State University
    Box 7619
    Raleigh, NC   27695-7619 USA

Dr. Robert Sanford
    Natural Resource Ecology Lab
    Department of Forest and Wood Sciences
    Colorado State University
    Fort Collins, CO   80523 USA

Dr. M. Scott Smith
    Department of Agronomy
    University of Kentucky
    Lexington, KY   40546 USA

Dr. Phillip Sollins
    School of Forestry and Environmental Studies
    Yale University
    New Haven, CT   06511 USA

Dr. F. J. Stevenson
    Department of Agronomy
    1102 South Goodwin Avenue
    University of Illinois
    Urbana, IL   61801 USA

Dr. J. W. B. Stewart
    Department of Soil Science
    University of Saskatchewan
    Saskatoon, Canada   S7N 0W0

Dr. Larry T. Szott
North Carolina State University
Tropical Soil Program
Yurimaguas, Loreto, Peru

Dr. Kevin R. Tate
Division of Land and Soil Sciences - DSIR
Private Bag
Lower Hutt
New Zealand

Dr. Robert L. Tate
Department of Soils and Crops
Rutgers University
New Brunswick, NJ   08903 USA

Dr. B. K. G. Theng
Division of Land and Soil Sciences - DSIR
Private Bag
Lower Hutt, New Zealand

Dr. Holm Tiessen
Department of Soil Science
University of Saskatchewan
Saskatoon, Canada   S7N 0W0

Dr. Goro Uehara
IBSNAT Project
University of Hawaii
1910 East-West Road
Honolulu, HI   96822 USA

Dr. Eric D. Vance
Institute of Arctic Biology
University of Alaska
Fairbanks, AK   99775 USA

Dr. J. A. van Veen
ITAL Box 48
6700AA
Wagenningen, The Netherlands

Dr. Peter Vitousek

Department of Biological Sciences
Stanford University
Stanford, CA   94305   USA

Dr. Koji Wada

Department of Agricultural Chemistry
Kyushu University 46
Fukuoka, 812   Japan

Dr. Russell Yost

Department of Agronomy and Soil Science
University of Hawaii
1910 East-West Road
Honolulu, HI   96822   USA

Preface

Poverty has its roots in the soil. Most of the poorer nations of the world are located in the tropics, in areas dominated by poor soils, unfavorable climates, or both. The impoverished and stressful state of many soils in the tropics stems from their high degree of weathering. Soil minerals which normally hold plant nutrients have been eroded to such an extent that they are no longer capable of retaining added nutrients. Consequently, highly leached tropical soils have high levels of acidity and toxic metals and low amounts of available nitrogen, phosphorus and the bases.

It is generally agreed that organic matter can play an important function in alleviating nutrient and stress problems in soils. But, very little is known about the specific biological processes involved in transformation of soil organic matter (SOM) and organic matter inputs into the ecosystem. This is particularly true of those tropical soils dominated by variable-charge clays and subjected to heavy rainfall over a long period of time.

This book is the result of a process that began with a questionnaire in July of 1986 and ended with a workshop in October 1988. The primary purpose of the whole effort was to produce a book; a book that attempted to identify gaps in knowledge, prescribe corrective measures and formulate research priorities related to ecological interactions that regulate organic matter dynamics in tropical ecosystems. Another important objective was to bring together scientists from diverse disciplines and provide them the opportunity to interact and "brainstorm" on controversial issues related to quantity and quality of organic matter in tropical soils and the role it might play in restoring fertility and productivity of agricultural and forestry systems in the tropics.

The idea for the format of this event came from a similar process that resulted in the publication of "Crop Productivity-Research Imperatives Revisited," an international conference held at Boyne Highlands Inn, October 13-18, 1985, edited by M. Gibbs and C. Carlson. The present book was strongly influenced by this format. From responses we received to questionnaires about tropical SOM which were sent to over 100 scientists, seven topics were selected as the major themes for this book. Teams of experts were commissioned to write a "position paper" on each of the seven themes. Authors were asked to be deliberately controversial and provocative. Each "position paper" was then submitted for evaluation to a separate panel of experts from such diverse fields as ecology, chemistry, physics, soil science and agriculture. Written reports from panel members were forwarded to the authors a few week before the workshop. During a week-long workshop on Maui, October 9-14, 1988, the authors and the reviewers gathered together to defend their positions, reconcile differences and produce "consensus papers" that have become the seven chapters in this book.

The workshop began with plenary sessions on the first day when the authors presented a summary of their position papers. Authors of each theme then met with their respective panels to modify and re-write the new sections. At the end of each day, panel reporters gave a short summary of the day's accomplishments. Three word processors and typists were at hand to make the changes and distribute the revised statements to the participants on the same day.

Each panel was also challenged with one or more paradigms related to their themes and asked to develop statements on research imperatives that would address the specific paradigms. For example, the statement "tropical soils are substantially lower in organic matter than temperate soils" was the challenge presented to the whole workshop and became the subject of heated debate and considerable caucusing during the workshop.

It was a pleasant task, and an educational experience, for me to be given the responsibility of organizing this event. It was made pleasant because of the hard work and dedication of numerous individuals that helped me. It became educational because of the caliber and qualifications of the many scholars that I dealt with during chapter preparations and review processes. The editors, Coleman, Oades and Uehara, have been gracious, patient and professional. I am personally indebted to David Coleman for all of the assistance and guidance he provided during proposal development, author and panelist assignments and the final preparation of this book. Cynthia Garver, of Callico Associates, served effectively as the "gadfly" during the workshop and expertly as the copy-editor when the chapters were received.

The monumental task of producing and publishing this book has been the responsibility of Princess Ferguson, Head of NifTAL's publication section. With a great deal of cooperation and assistance from Jan Heavenridge of the University of Hawaii Press, she was able to adhere to a very tight publication schedule.

Last, but not least, I would like to acknowledge the support of the staff of the NifTAL Project during this whole process. In particular, I am indebted to Susan Tomoso for her good-natured dedication as Workshop Coordinator, Paul Woomer (and Dr. H. Ikawa, University of Hawaii), Kevin Keane and the field crew for field-trip preparations, Karean Zukeran for local arrangements, Susan Hiraoka and Rani Machoi for typing and proof-reading, Sally Ekdahl for final editing assistance, and to Anna Gillies for art work. I am also indebted to Dr. Samir El-Swaify for presenting the opening "pep talk" and moderating the closing "wrap-up" discussions on "Opportunites for International Collaboration."

Support for this endeavor was provided primarily by the National Science Foundation, Ecology Program. Several projects of the US Agency for International Development's Office of Agriculture namely IBSNAT, NifTAL and Trop Soils Projects, provided travel assistance to participants from developing countries.

B. Ben Bohlool
Organizer
NifTAL Project
University of Hawai
1000 Holomua Avenue
Paia, HI 96779 USA
April 1989

Soils of the Tropics

LEGENDS — Orders and Suborders*

ALFISOLS | A 2 | Udalfs
| A 3 | Ustalfs

ARIDISOLS | D |

ENTISOLS | E 1 | Aquents
| E 3 | Psamments

INCEPTISOLS | I 2 | Aquepts

ULTISOLS | U 1 | Aquults
| U 2 | Humults
| U 3 | Udults
| U 4 | Ustults

INCEPTISOLS | I 4 | Tropepts

MOLLISOLS | M 1 | Albolls
| M 3 | Rendolls
| M 5 | Ustolls

OXISOLS | O 1 | Orthox
| O 2 | Ustox

VERTISOLS | V 2 | Usterts

OTHER | X | Mountain areas
| | Lakes

Adapted from a 1971 manuscript soil map prepared by the Soil Geography Unit, Soil Conservation Service, U.S. Department of Agriculture.
*Third character in mapping symbol refers to inclusions in the suborders, which are identified in the accompanying legend.
The representation of international boundaries on this map is not necessarily authoritative.

DISTRIBUTION OF ORDERS AND
PRINCIPAL SUBORDERS AND GREAT GROUPS

In this legend, only general definitions for orders and suborders are given. For complete definitions of these, and for the great groups comprising some of the mapping units, see *Soil Classification, A Comprehensive System, 7th Approximation,* Soil Conservation Service, U.S. Department of Agriculture, 1960, and the March 1967 Supplement; also *Soil Taxonomy of the National Cooperative Soil Survey,* Soil Conservation Service, U.S. Department of Agriculture.

A ALFISOLS Soils with subsurface horizons of clay accumulation and medium to high base supply; either usually moist* or moist for 90 consecutive days during a period when temperature is suitable for plant growth
 A 2 *Udalfs* temperate to hot, usually moist
  A 2e with Troporthents
 A 3 *Ustalfs* temperate to hot, dry more than 90 cumulative days during periods when temperature is suitable for plant growth
  A 3a with Tropepts
  A 3b with Troporthents
  A 3c with Tropustults
  A 3d with Usterts
  A 3e with Ustochrepts
  A 3f with Ustolls
  A 3g with Ustorthents
  A 3h with Ustox
  A 3j Plinthustalfs with Ustorthents

D ARIDISOLS Soils with pedogenic horizons, usually dry in all horizons and never moist as long as 90 consecutive days during a period when temperature is suitable for plant growth

E ENTISOLS Soils without pedogenic horizons
 E 1 *Aquents* seasonally or perennially wet
  E 1b Psammaquents with Haplaquents
  E 1c Tropaquents with Hydraquents
 E 3 *Psamments* sand or loamy sand textures
  E 3b with Orthox
  E 3d with Ustalfs
  E 3e with Ustox

I INCEPTISOLS Soils with pedogenic horizons of alteration or concentration but without accumulations of translocated materials other than carbonates or silica; usually moist or moist for 90 consecutive days during a period when temperature is suitable for plant growth
 I 2 *Aquepts* seasonally wet
  I 2c Haplaquepts with Humaquepts
  I 2e Humaquepts with Psamments
  I 2f Tropaquepts with Hydraquents
  I 2g Tropaquepts with Plinthaquults
  I 2h Tropaquepts with Tropaquents
  I 2j Tropaquepts with Tropudults
 I 4 *Tropepts* continuously warm or hot
  I 4a with Ustalfs
  I 4b with Tropudults
  I 4c with Ustox

*Soil moisture terms: Dry—soil moisture below permanent wilting
        Moist—soil moisture above permanent wilting
        Wet—soil saturated

M MOLLISOLS Soils with nearly black, humus-rich surface horizons and
high base supply; either usually moist or usually dry
   M 1 *Albolls* light gray subsurface, seasonally wet
      M 1a with Aquepts
   M 3 *Rendolls* subsurface has much calcium carbonate but no clay
accumulation
      M 3a with Usterts
   M 5 *Ustolls* temperate to hot, dry more than 90 cumulative days in year
      M 5a with Argialbolls
      M 5c with Usterts

O OXISOLS Soils with pedogenic horizons that are mixtures principally of
kaolin, hydrated oxides, and quartz, and are low in weatherable minerals
   O 1 *Orthox* hot, nearly always moist
      O 1a with Plinthaquults
      O 1b with Tropudults
   O 2 *Ustox* warm or hot, dry for long periods but moist more than 90
consecutive days in the year
      O 2a with Plinthaquults
      O 2b with Tropustults
      O 2c with Ustalfs

U ULTISOLS Soils with subsurface horizons of clay accumulation and low
base supply, usually moist or moist for 90 consecutive days during a period
when temperature is suitable for plant growth
   U 1 *Aquults* seasonally wet
      U 1b Plinthaquults with Orthox
      U 1c Plinthaquults with Plinthaquox
      U 1d Plinthaquults with Tropaquepts
   U 2 *Humults* temperate or warm and moist all of the year, high content
of organic matter
      U 2a with Umbrepts
   U 3 *Udults* temperate to hot, never dry more than 90 cumulative days
in the year
      U 3d Hapludults with Dystrochrepts
      U 3e Rhodudults with Udalfs
      U 3f Tropudults with Aquults
      U 3g Tropudults with Hydraquents
      U 3h Tropudults with Orthox
      U 3j Tropudults with Tropepts
      U 3k Tropudults with Tropudalfs
   U 4 *Ustults* warm or hot, dry more than 90 cumulative days in the year
      U 4a with Ustochrepts
      U 4b Plinthustults with Ustorthents
      U 4c Rhodustults with Ustalfs
      U 4d Tropustults with Tropaquepts
      U 4e Tropustults with Ustalfs

V VERTISOLS Soils with high content of swelling clays; deep, wide cracks
develop during dry periods
   V 2 *Usterts* cracks open more than 90 cumulative days in the year
      V 2a with Tropaquepts
      V 2b with Tropofluvents
      V 2c with Ustalfs

X SOILS IN AREAS WITH MOUNTAINS Soils with various moisture and
temperature regimes; many steep slopes, relief and total evaluation vary
greatly from place to place. Soils vary greatly within short distances and
with changes in altitude; vertical zonation common.

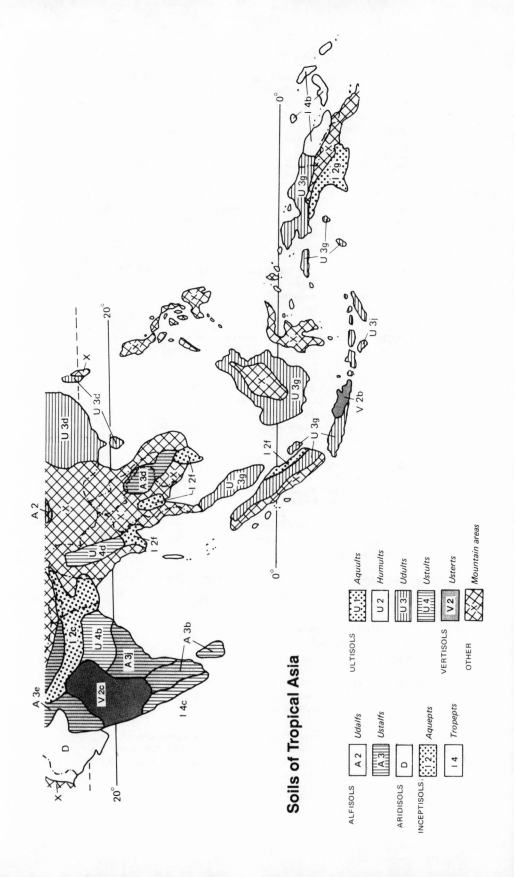

Soils of Tropical Asia

ALFISOLS

A 2	*Udalfs*
A 3	*Ustalfs*

ARIDISOLS

D	

INCEPTISOLS:

I 2	*Aquepts*
I 4	*Tropepts*

ULTISOLS

U 1	*Aquults*
U 2	*Humults*
U 3	*Udults*
U 4	*Ustults*

VERTISOLS

V 2	*Usterts*

OTHER

	Mountain areas

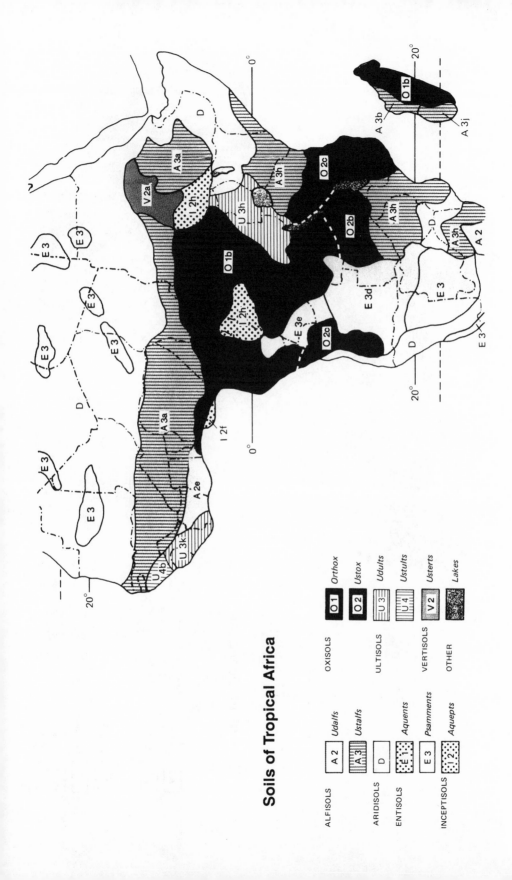

Soils of Tropical Africa

ALFISOLS			OXISOLS		
A 2	*Udalfs*		O 1	*Orthox*	
A 3	*Ustalfs*		O 2	*Ustox*	
ARIDISOLS			ULTISOLS		
D			U 3	*Udults*	
ENTISOLS			U 4	*Ustults*	
E 1	*Aquents*		VERTISOLS		
E 3	*Psamments*		V 2	*Usterts*	
INCEPTISOLS			OTHER		
I 2	*Aquepts*			*Lakes*	

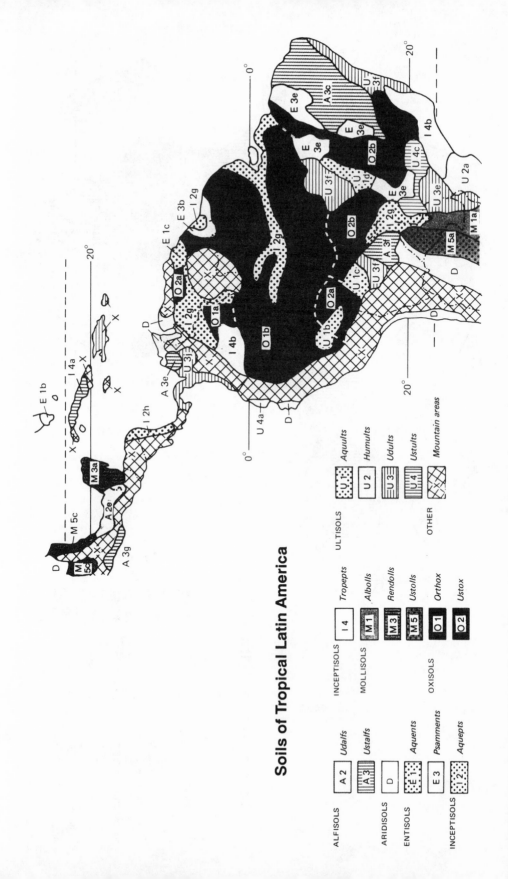

Soils of Tropical Latin America

ALFISOLS			INCEPTISOLS	I 4	Tropepts		ULTISOLS			
A 2		Udalfs					U 1		Aquults	
A 3		Ustalfs	MOLLISOLS	M 1	Albolls		U 2		Humults	
ARIDISOLS				M 3	Rendolls		U 3		Udults	
D				M 5	Ustolls		U 4		Ustults	
ENTISOLS			OXISOLS	O 1	Orthox		OTHER		Mountain areas	
E 1		Aquents		O 2	Ustox					
E 3		Psamments								
INCEPTISOLS	I 2	Aquepts								

Introduction

Soil Organic Matter:
Tropical vs. Temperate

Why should we need a book specifically on organic matter in soils of tropical regions? Soils in the tropics range widely from very young fertile soils, developed for example on volcanic ash, i.e., Andisols, to old infertile soils, i.e., Ultisols. Water regimes vary from arid to continuously wet. We also know that differences in the amounts and nature of organic materials vary as much over small distances as they do from temperate zones to the tropics. In addition, the indications are that the decomposition of plant materials is dominated by similar groups of organisms in the decomposer cycle across the world. What then is the difference?

As a first approximation we can define the tropics as the area between latitudes 23.5° North and South. These lines need to be bent here and there on continental masses and around features such as the Himalayas. Annual rainfall can be extremely diverse, varying from about zero to 10,000 mm. The major consistent difference between temperate and tropical regions is temperature. On average, the tropics are 15°C warmer than temperate regions and in addition have no cold winters. This higher temperature throughout the year results in a several-fold increase in all chemical and biochemical reaction rates. This has enormous implications for the soil development and biogeochemical cycling in tropical ecosystems. Providing water is present — which is the case in humid tropics with rainfalls from 3,000 to 6,000 mm per year — rates of primary production and decomposition are several times more rapid than in temperate regions. This does not mean that organic matter contents of tropical soils are greater or less than in soils of temperate regions or that the chemistry is different; it does mean that the turnover of organic matter is more rapid. This has a major influence on the cycling of the range of organogenic elements in tropical ecosystems. For example, during rapid plant growth there is a great demand for essential nutrients, while rapid decomposition may lead to substantial nutrient release.

In natural ecosystems, some near-equilibrium exists which takes account of these factors. When this equilibrium is disturbed by clearing forests and agricultural practices, it is vitally important to understand the impact of the changed conditions on the cycling of carbon, nitrogen, phosphorus, and other biogenic elements. Thus, particularly in the tropics, man's ignorant intervention in a naturally balanced ecosystem in the quest for food and in many cases a desire for monetary gain may lead to a very rapid destruction of the environment.

One characteristic of some tropical soils resulting from rapid reaction rates is deep weathering and intensive leaching as in Oxisols and Ultisols. The low charge minerals and oxides which result are not efficient in holding nutrient cations in the root zone of vegetation. Thus, organic matter becomes a vitally important sink and source of essential elements and must, therefore, be conserved even more than in temperate regions. Environments like these occur mainly in the humid belt near the equator. The rainforests of Indonesia and the Amazon and the Congo river basins are prime examples.

There are two other very important reasons why organic matter in tropical soils is so important. The tropics are covered largely by developing countries inhabited by about one half of the world's population. Moreover, it is in these countries where the largest increases in population are expected and where demands for food production will be greatest.

In the past we have assumed that how an individual country managed its soil and forest resources was its own business. This assumption is now being replaced by a new sense of global awareness in which people are beginning to realize that phenomena on as large a scale as global climate can be changed as a consequence of human activities. For example, the concentration of CO_2 in the atmosphere has increased from about 280 ppm to 340 ppm since the Industrial Revolution due to fossil fuel burning, deforestation, and agriculture. The rate of increase remains about 1-5 ppm CO_2 per annum.

Is there a difference between soil organic matter of the tropics and temperate regions? Perhaps the diversity of conditions in the tropics overwhelm any major differences that may exist between tropical and temperate areas and preclude answering this question at this time. The authors of the first chapter examine this issue and conclude that no special quality of the organic constituents sets those from the tropics apart from their temperate region counterparts. The authors of the second chapter note that in general the accumulation and distribution of soil organic matter and its nutrient supply depend on its rates of replenishment and decomposition, and on the capacity of the soil to protect humic substances from microbial attack. If the stability of organic matter largely results from protective processes in soils rather than from the creation of recalcitrant chemical forms, what is the mechanism of this protective process?

In Chapter 3, some evidence is provided to explain how soil minerals protect organic matter and why a soil is more than the sum of its inorganic and organic parts. Interactions of organic molecules with iron and aluminum sesquioxides bring about deep-seated changes in soil physicochemical properties.

In addition to these chemical changes, biological processes regulate organic matter dynamics. Termites and earthworms, for example, contribute to the dynamics of soil organic matter by influencing metabolic processes, by affecting litter input and decomposition, and by aggregating tiny particles into larger ones to alter soil structure. Microorganisms and plant roots also affect the physics of soils by forming sinks for water, nutrients, and oxygen. In describing the above processes, the authors of Chapter 4 hypothesize that soil organic matter dynamics are essentially the same in temperate and tropical systems.

In the fifth chapter, the authors consider the fate of organic residues when they are mixed into a soil. Their aim was to capture, organize, and condense what is known about this subject to enable users of this knowledge to predict and control outcomes of organic residue management. This chapter clearly demonstrates the need to organize our knowledge into simulation models. The latter task was assigned to the authors of Chapter 6 and this "modeling group" drew on information provided by authors of the earlier chapters and addressed the issue of universality of patterns and processes, thereby diffusing the need to distinguish between tropical and temperate systems. The principal outcome seemed to be increased desire among researchers to modify and expand the CENTURY model to mimic processes and predict soil organic matter content in a wide range of environmental situations. To do so, users of the model will need standard methods for assessing the quantity and quality of soil organic matter. Thus, the authors of the last chapter outline suitable techniques to characterize soil organic matter. The techniques demand that the investigator be aware of the physical nature of the soil and the ways in which physical fractions can be subsequently characterized chemically. With an increasing concern for sustainable ecosystem status and productivity, the authors of the last chapter particularly emphasize biologically significant soil organic matter pools as the basis for characterizing soil organic matter.

In the last several years, a large group of scientists worldwide have been interested in fostering research on basic processes occurring in tropical ecosystems. Within North America these efforts have been supported by several agencies, primarily the National Science Foundation (NSF) and the Agency for International Development (AID). A workshop was held in Athens, Georgia, in November, 1985, on the topic: "Tropical Soil Biology" (Hayes and Cooley, 1987). This led to further research interest in temperate/tropical comparisons of terrestrial carbon, nitrogen, sulfur, and phosphorus cycling (also funded by NSF) and the publication of a special issue of **Biogeochemistry** 5:(1) (1988) on this topic.

With the encouragement of Dr. Patrick W. Flanagan and others in NSF, a group of us (Bohlool, Coleman, and Uehara) submitted a proposal and organized an intensive five-day workshop on "Dynamics of Tropical Soil Organic Matter" (TSOM), which was held in Kahului, Maui, Hawaii, October 9 to 14, 1988. Our objectives were both straightforward and ambitious: to bring together a group of world recognized authorities in organic matter studies, and to ascertain the principal areas where gaps remain in our knowledge. We endeavored to bring scientists from several developing countries in tropical regions. We were able to entice only a few participants from developing countries to our meeting, perhaps due to the time and effort involved. Therefore, we felt it essential to bring together an effective synthesis of seven major interest areas in a volume which would be timely and affordable. Thanks to the University of Hawaii Press, and the considerable efforts of our organizer, B. Ben Bohlool, we have achieved both of these goals. We hope the readers find (at least some of) these chapters of interest as they delve into the intricacies of the fascinating world of organic matter in tropical ecosystems.

We found the meeting on Maui strenuous, invigorating, and frustrating as we endeavored to stretch people's thoughts and concepts across a wide spectrum of biological, chemical, and physical aspects of TSOM. If only a portion of the excitement and interest which we felt comes through in this volume, we will feel that the efforts involved in holding the workshop and writing the book will have been worthwhile.

D. C. Coleman
J. M. Oades
G. Uehara

Chapter 1

Constituents of Organic Matter in Temperate and Tropical Soils

Benny K. G. Theng, Kevin R. Tate, and Phillip Sollins

with Norhayati Moris, Nalini Nadkarni, and
Robert L.Tate III

Abstract
The organic constituents of tropical soils do not possess any special
qualities that set them apart from those of temperate soils. The study of
soil organic matter (SOM) constitution in temperate and tropical soils
may therefore be approached using similar procedures. Such procedures (physical fractionation and chemical extraction, for example)
should truly reflect the dynamic nature of the constituents: that is, they
should be biologically meaningful. Accordingly, the constituents of SOM
are discussed here in terms of their relative stability in soil, from the labile
through the stable forms.

Litter is included in the labile component; other constituents in this
category are plant roots, organic plant nutrients, macroorganisms
(fauna), microorganisms, macroorganic matter or light fraction, water-
soluble forms, and nonhumic substances. The stable component of SOM
is largely identifiable with humic substances. Although humic substances

and other structurally complex polymers have some intrinsic resistance against microbial attack, their stability in soil is mainly due to interactions with the mineral constituents and entrapment within clay aggregates.

The comminution of litter by macroorganisms and the transformations of organic matter in soil by microorganisms follow similar pathways for both temperate and tropical ecosystems. The relative pool size of the various SOM constituents, however, is determined by climatic and edaphic factors. The amount and composition of these constituents must therefore be assessed along well-defined gradients of climate, soil mineralogy, and physics.

The interplay of environmental factors gives rise to some overlap and variability in organic matter contents of temperate and tropical soils. An example of this is shown in Figure 1, which compares soil organic C levels in

Figure 1. Relationship between organic carbon content and rainfall for Ultisols (Krasnozems) from tropical, and cool, temperate regions of Australia. The overall importance of the environmental factors determining soil carbon contents is: rainfall > pH > clay content > temperature (for tropical regions) and rainfall > pH > temperature > clay content (for temperate regions). (Redrawn from Spain et al., 1983, with permission from the publisher.)

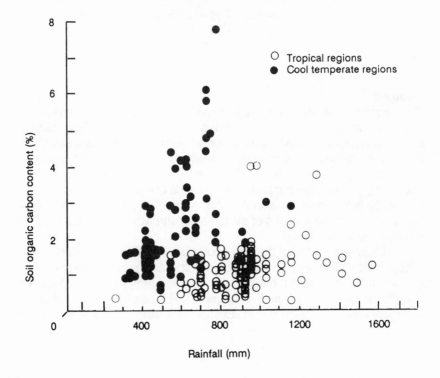

Dynamics of SOM in Tropical Ecosystems

Ultisols in the tropical regions with those in the cool, temperate regions of Australia (Spain et al., 1983). In this instance, rainfall was the major factor that determined the C level. Soil pH, temperature and clay content also influenced organic C contents, but these factors apparently contributed more in tropical than in temperate soils. This feature is also shown in Figure 2 for plant litter and soil organic matter; a wider range of SOM values has been found for tropical rainforests than between these and the forests of temperate and boreal regions.

Figure 2. Organic matter in surface litter and in soil of lowland forests ranging from the Arctic Circle to near the equator. (Redrawn from J.M. Anderson and Swift, 1983, with permission from the publisher.)

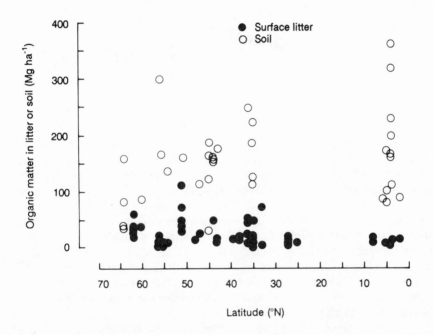

It is therefore essential to understand the causes underlying differences in the amount and composition of SOM across soil types and, if possible, to predict how these parameters might change with differing inputs or management. To do this, we must study soils that differ in only one factor, or at most a few factors, such as climate, vegetation, mineralogy, or aeration. The emphasis to date has been on the effects of climate and the results, although interesting, are not definitive. In the future the emphasis should be on understanding the variability of SOM within a climatic region (Figure 3).

Figure 3. Relationship between total soil organic carbon and the temperature to precipitation ratio (T/P) for a number of sites in six tropical and subtropical ecosystems or "life zones" (Holdridge, 1967). These include tropical wet and rain forest, subtropical and tropical moist forest, and subtropical and tropical dry forest (Redrawn from Brown and Lugo, 1982, with permission from the publisher.)

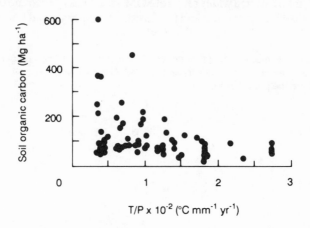

The overlap and variability in quantity of organic C in climatic regions also extend to the various constituents of SOM (e.g., Kononova, 1975). Several questions arise. First, are these constituents qualitatively similar in temperate and tropical soils? In an earlier review, K. R. Tate and Theng (1980) argued that this is indeed the case since the precursors of (dead) SOM and the biochemical pathways involved in its formation are similar. Second, do tropical soils contain a larger proportion of recalcitrant constituents than do soils in temperate regions? This would imply that the content of labile constituents in tropical soils is relatively low, or that recalcitrant compounds should make up a large proportion of the total. Third, is the biostability or recalcitrance of some constituents intrinsic to their chemical (molecular) structure, or is it due to interactions with clays and minerals?

Our objectives are to provide answers to these questions in the light of the available data and to suggest research imperatives.

The extensive literature on organic matter constituents, particularly with respect to temperate soils, has made it necessary to be selective as to what aspects, and how much of a given topic, are to be included in a chapter of limited size. The chemistry of humic substances, for example, is not discussed in detail. Furthermore, we have attempted to treat the various constituents from a dynamic perspective, stressing their interactions with one another and with the mineral fraction of soil.

Dynamics of SOM in Tropical Ecosystems

Accordingly, we have included such topics as litter, fauna, microorganisms, and macroorganic matter. That the section on humic substances occupies a disproportionate amount of space is because so much more is known about them than is the case for the other soil organic constituents.

Concepts and Definitions
Living and Dead Soil Organic Matter

To facilitate this discussion, we divide SOM into living and nonliving, or dead, components (Theng, 1987). The living component, which rarely makes up more than 4% of the total soil organic C, may be subdivided into three compartments: plant roots (5-10%), macroorganisms or fauna (15-30%), and microorganisms (60-80%). The values in parentheses denote relative pool sizes. Although roots comprise a minor constituent of the living component, their passage and distribution through soil may have a greater influence on soil processes than their pool size suggests. The nonliving component, making up as much as 98% of the total soil organic C, may be subdivided into macroorganic matter and humus.

Macroorganic matter is often the smaller of the two nonliving compartments, commonly containing 10-30% of the total soil organic C. It consists largely of plant residues in varying stages of decomposition, capable of being retained on a 250-μm sieve. Macroorganic matter can be equated with the "light fraction" obtained by flotation on liquids with a density of 1.6-2.0 g cm^{-3}. In allophanic soils, much light fraction material is substantially smaller than 250-μm. If much pumice is present in these soils, a density approaching 1.2 g cm^{-3} is preferable since then the majority of the pumice sinks (Boone et al., 1988).

The dead organic matter remaining after separation of the macroorganic matter or light fraction is commonly referred to as humus and consists of nonhumic and humic substances. The former constituents make up about 30% of humus and comprise well-defined classes of organic compounds, such as carbohydrates, lipids, organic acids, pigments, and proteins. Humic substances, which make up the bulk of dead SOM, have been the most intensively studied. Because of their molecular complexity, heterogeneity, and polydispersity, however, we are still unable to assign chemical structures to these materials. Three categories or fractions have conventionally been recognized: fulvic acid, which is soluble in acid and alkali; humic acid, which is only soluble in alkali; and humin, which is not soluble in either medium. These definitions are merely operational in that no sharp boundary exists between these fractions in terms of physico-chemical properties. The real drawback with this fractionation scheme, however, is the lack of relation to dynamic processes.

Litter also consists of dead plant (and some animal) matter. Because it lies on the soil surface, strictly speaking litter is not a constituent of SOM.

Because of its importance in nutrient cycling and humus formation, however, the litter layer is considered an integral part of the soil profile.

The Dynamic Perspective

SOM may also be divided into labile and stable compartments, or fractions. These terms are relative concepts based on the rate of decomposition in soil of a particular constituent. As such, they are more closely related to function and mechanism than are the definitions of fulvic acid, humic acid and humin, which are essentially chemical and operational.

A number of models have been proposed to describe SOM dynamics (Paul and van Veen, 1978; van Veen et al., 1984; Pastor and Post, 1986; Jenkinson et al., 1987; Parton et al., 1987, 1988). Although the results vary in detail, depending on the underlying assumptions, it is generally accepted that labile constituents decompose within a few weeks or months whereas their stable counterparts may persist in soil for years or even decades. Included among the labile constituents are comminuted plant litter, macroorganic matter or light fraction, the living component or biomass, and nonhumic substances that are not bound to the mineral constituents. The stable organic constituents in soil include humic substances and other organic macromolecules that are intrinsically resistant against microbial attack, or which are physically protected by adsorption on mineral surfaces or entrapment within clay and mineral aggregates.

To a greater or lesser extent, the various organic constituents in soil interact with one another and with the mineral fraction. For the sake of clarity, however, we shall describe each constituent separately, from the labile fractions to the stable forms of SOM.

Constituents of Soil Organic Matter

Litter — Precursor to or Component of SOM?

Relatively much is known about litter behavior in forest ecosystems, at least in terms of its accumulation on, and disappearance from, the soil surface (Brown and Lugo, 1982; J. M. Anderson and Swift, 1983; Spain and Hutson, 1983). Figure 4 summarizes some of the data for tropical rainforests and temperate deciduous forests, while Figure 5 shows the effect of macroclimate on total litterfall in some tropical and subtropical ecosystems. As might be expected, the decomposition rate constant (k_L) for litter in tropical forests is generally greater than one, corresponding to a turnover time of less than 1 year. In contrast, turnover times in temperate forests are commonly between 1 and 2 years. However, there is also much overlap in k_L values, indicating that the rate of decay for litter (of comparable quality) in the tropics need not always be greater than in temperate regions.

Dynamics of SOM in Tropical Ecosystems

Figure 4. Relationship between total litterfall and litter standing crop in tropical rainforests and temperate deciduous forests. The ratio of litterfall to litter standing crop (content) is the litter decomposition rate constant denoted by the symbol k_L. The turnover time is the reciprocal of k_L. The k_L values of 0.5, 1.0, and 2.0 years are shown. (Redrawn from J.M. Anderson and Swift, 1983, with permission from the publisher.)

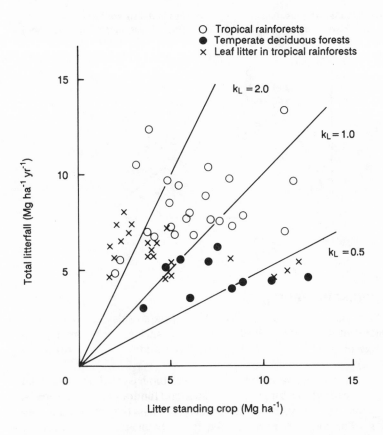

The influence of litter quality on degradability also becomes apparent when the data for leaf litter are compared with those for total litter, which includes 30-40% woody materials with a relatively high lignin content. Indeed, for a range of forest ecosystems both initial lignin and the C/N ratio of leaf litter are well correlated with decomposition rate (Meentemeyer, 1978; Laishram and Yadava, 1988). For any given site, however, litter decay rates can be related to the abundance, activity, and composition of soil organisms (Dietz and Bottner, 1981).

The comminution and decomposition of litter increases its cation exchange capacity (Toutain, 1987). In addition to an increase in carboxyl-C, the alkyl-C content rises while that of both carbohydrate-C and aromatic-C decreases as decomposition progresses (Zech et al., 1985, 1987; Hempfling et al., 1987).

Figure 5. Total litterfall as a function of T/P ratio for sites in the six ecosystems referred to in Figure 3. (Redrawn from Brown and Lugo, 1982, with permission from the publisher.)

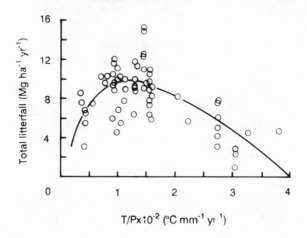

Macroorganisms (Fauna)

Soil invertebrates are conventionally divided into three size classes: microfauna, which include protozoa and nematoda; mesofauna, such as enchytraeidae and acari; and macrofauna, exemplified by earthworms and termites (Luxton and Petersen, 1982).

Figure 6 shows the variation in soil fauna biomass with mean annual temperature for ecosystems ranging from the arctic tundra to tropical forests. Tropical savannas and forests have a lower fauna biomass than their temperate counterparts. For forest ecosystems, the fauna biomass is often inversely related to the accumulated weight of organic matter on the soil surface, but no clear relationship exists between fauna biomass and annual litterfall.

Climate also influences the composition of macroorganisms. In exposed sites of temperate alpine and subalpine regions, for example, the biomass is largely made up of mesofauna, whereas in temperate and tropical lowland forests it is dominated by the macrofauna (Petersen, 1982).

Although some soil invertebrates can digest lignin and polysaccharides of plant origin and so act as primary decomposers, macroorganisms contribute much less than microorganisms to the total soil respiration. There is little doubt, however, that soil fauna play a primary role in comminuting surface litter,

Dynamics of SOM in Tropical Ecosystems

incorporating the fragmented products into the soil, and converting it into excreta. Generally, 20-40% of the annual litterfall is processed in this way (Kitazawa, 1967). Besides facilitating and accelerating the decomposition of plant residues by microorganisms, soil fauna contribute to nutrient cycling by grazing on the microflora as well as by mixing soil materials within the profile.

Earthworms and termites, in particular, are efficient soil homogenizers in both temperate and tropical soils (Herbillon, 1980; Lee, 1983). At some sites, termites may play a critical role in bringing weatherable minerals to the surface from deeper horizons, thereby influencing soil properties (Pomeroy, 1983). They have even been suggested as causal agents for stone lines in highly weathered soils (Wielemaker, 1984).

The interaction and symbiosis between macro- and microorganisms in comparable soils of the tropical and temperate regions, rather than their separate activity in decomposing plant materials and in nutrient cycling, warrant further study.

Figure 6. Relationship between the total soil fauna biomass and the mean annual temperature, ranging from the alpine to the tropical zones. (Redrawn from Petersen, 1982, with permission from the publisher.)

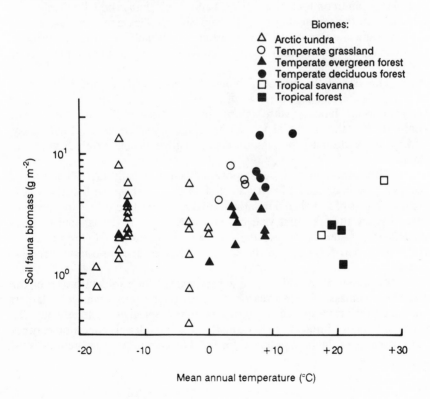

Microorganisms

Although there are still problems associated with the measurement of soil microbial biomass, the amount of C contained in the biomass mostly ranges from 1 to 3% of the total soil organic C. The C/N ratio of the microbial biomass ranges between 5:1 and 15:1, depending on whether the microbial population is dominated by bacteria or fungi, respectively. The latter ratio, or something approaching it, is more likely in acid soils, especially if they are well-drained (R. L. Tate, 1987).

Figure 7 shows the relationship between microbial C and total C for a number of soils from temperate and tropical regions. Values lying outside the two demarkation lines are soils in which pH and other soil factors (such as moisture) affect the results.

Little information is available on biomass contents of tropical soils. The limited data suggest low biomass levels in these soils. These values suggest a rapid turnover of biomass, especially when moisture does not limit decomposition. In contrast, soils derived from volcanic ash give high biomass contents, consistent with the large accumulation of organic matter in these soils. This may be ascribed to complex formation involving aluminum and iron.

Zunino et al. (1982) showed that the decomposition of simple and complex organic compounds is slower in allophanic than in nonallophanic soils, especially in the initial stages of decomposition. This might be because allophane has a large propensity for adsorbing and immobilizing organic substances (Theng, 1987). In addition, the biomass itself and its metabolites may be stabilized by allophane.

The size of the microbial biomass is not always a good indicator of microbial activity in that a range of situations from high biomass with low turnover to low biomass with high turnover is possible. Factors such as the composition of the microbial population, frequency and quality of food supply, and an active predator population could singly and collectively have an influence.

The formation and decomposition of isotopically labeled microbial biomass have been studied by Ladd et al. (1981). During the first few weeks turnover times of less than 6 months were observed; these values increased to several years after the first year of decomposition. As the labeled biomass decomposes faster than does the nonliving component, the proportion of residual labeled C (and N) in the soil decreases as decomposition progresses.

A supply of available C is a prerequisite for rapid turnover of the microbial biomass. For this reason, the activity of microorganisms is largely confined to the rhizosphere. However, in some tropical ecosystems where the soil receives large inputs of organic matter and rainfall, this condition may not apply because of the high potential for SOM leaching. Measurement of SOM turnover and microbial populations in nonrhizosphere and rhizosphere soils of such ecosystems seems warranted.

Figure 7. Relationship between microbial biomass carbon and soil organic carbon for a range of temperate and tropical soils. The limits at which the ratio of C_b to C_s equals 0.01 and 0.03 are shown. The biomass values for soils from England, Scotland, Nigeria, and India were obtained from the flush of carbon mineralization after chloroform fumigation. The values for New Zealand soils were estimated using chloroform fumigation and substrate-induced respiration; climosequence of soils in tussock grassland (cf. Fig (8) (Δ); and soils derived from volcanic ash (+). Compiled from data of Jenkinson et al., 1976; Jenkinson and Ladd, 1981; Sparling, 1981; Ross et al., 1980; Ross and K.R. Tate, 1982; Dash et al., 1985; Sparling and West, 1988a, 1988b.)

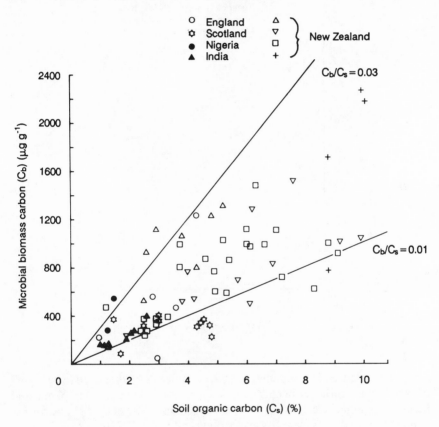

Macroorganic Matter or Light Fraction

The amounts of macroorganic matter and the propensity for its decomposition vary widely among soils. Acid soils in cold and dry regions, for example, frequently contain appreciable amounts of macroorganic matter because low pH and temperature, combined with water stress, limit litter comminution by soil fauna (Stout and Lee, 1980). The quality of plant litter (see Figure 4) and the aeration, mineralogy, and nutrient status of the soil are additional factors that affect macroorganic matter contents. Accordingly, variations in contents for tropical soils are likely to be as wide as those for temperate soils (Figure 8).

Figure 8. Variation with climate of the contents of organic matter in the light fraction of a range of temperate and tropical soils. Broken curve indicates hypothetical relationship. T = annual temperature (°C); P = annual precipitation (mm). (Compiled from data of Molloy and Speir, 1977, and Sollins et al., 1984.)

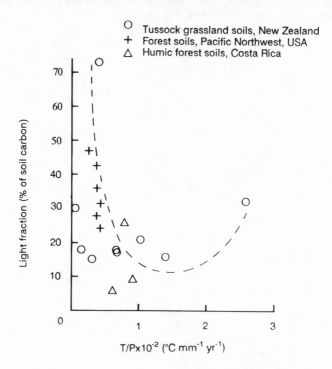

Knowledge of the amount of macroorganic matter is needed for studying organic matter turnover because of its intermediary role between plants and humus. As might be expected, the chemical composition of the light fraction is comparable to that of plant tissues (Molloy and Speir, 1977).

In temperate soils, varying proportions of the soil C and N are present in the light fraction (Ford et al., 1969; Perrott and Sarathchandra, 1987). This variability depends on the C/N ratio of the plant material from which the light fraction is derived (e.g., Sollins et al., 1984) and such factors as soil C content,

Dynamics of SOM in Tropical Ecosystems

pH, and the abundance of earthworms. Similar relationships should hold for tropical soils.

In keeping with its labile nature, the amounts of light fraction in soil vary markedly with season (Spycher et al., 1983). In both temperate and tropical soils, the labile character of the light fraction is also indicated by the disproportionately large contribution it makes to the plant-available N pool, especially if its C/N ratio is low (Sollins et al., 1984), and to C losses under cultivation. For example, in five tropical clay soils, the rate of loss of organic C from the light fraction was 2 to 11 times greater than that from the heavy (> $2\,g\,cm^{-3}$) fraction (Dalal and Mayer, 1986b). Physical inaccessibility to microorganisms of the organic matter in the latter situation is the most likely explanation for these differences in loss rate (Sollins et al., 1984).

Besides providing a useful insight into organic matter cycling, the light fraction also serves as an index for the consequences of changes in soil management in both temperate and tropical soils. For this reason, the light fraction should be considered separately from humus in studying organic matter dynamics.

Other Labile Forms

The labile forms of SOM also serve as a source of plant nutrients. The supply of nutrient elements to soil, from mineralization of organic forms, is particularly important to tropical agriculture (Tiessen et al., 1984b). Because of the low concentrations and inherent instability, however, the constitution of labile SOM remains largely unknown. In the case of organic P, the use of soil sequences has opened a window to this problem. Using alkali extracts of soils, K. R. Tate and Newman (1982), for example, observed that when biological activity is restricted by climate or soil conditions, potentially labile forms accumulate, largely as orthophosphate diesters of microbial origin. In temperate soils the amount of these compounds declines under cultivation (K. R. Tate, 1984). In tropical soils where biological processes are unrestricted, a small, steady supply of P may be derived from this source.

Soil microorganisms, as well as their metabolites, can also act as reservoirs of potentially available nutrients (e.g., McLaughlin et al., 1988) and carbon (R. L. Tate, 1987). Nutrients in this pool become available to plants if there is a decline in the overall soil biomass level as would occur when grasslands are converted to cultivation for crops.

As indicated earlier, the rapid turnover of labile organic forms of plant nutrients, including the microbial biomass, requires a supply of available C. This C is defined as soil organic carbon that heterotrophic microorganisms can readily use as an energy and C source (Davidson et al., 1987). A large amount of this carbon is water soluble and seldom exceeds 200 mg kg^{-1} in temperate soils. Probably much less is present in most tropical soils, although no information is available.

Available C originates from such sources as surface desorption, litter dissolution, exudation, plant root sloughing and exfoliation, microbial metabolites, and hydrolysis of more stable forms of organic matter (McGill et al., 1986). Little, if any, information is available on the relationship between water-soluble organic C and microbial biomass. Nor is much known about the fluxes of C through this water-soluble pool in temperate and tropical soils. It is worth noting that ^{13}C- and ^{1}H-NMR (solution and solid state) spectroscopy have been successfully used to characterize aquatic humic substances (e.g., Orem and Hatcher, 1987). To date, these techniques have not been applied to characterize available C from soil.

However, the role of water-soluble SOM in pedological processes has been extensively studied (e.g., Andreux et al., 1987). For example, water-soluble organic matter is implicated in the inverse relationship of the Si/Al ratio of soil leachates with temperature (Sollins et al., 1988). The rapid disappearance through leaching of this soluble organic matter in many tropical soils may effectively prevent Al chelation.

Nonhumic Substances

A wide range of nonhumic substances is likely to be represented in the labile organic fraction of all soils, including water-soluble organic matter (e.g., K. R. Tate and Theng, 1980). In general, the persistence in soil of these substances is inversely related to their molecular size and complexity (Haider and Martin, 1981). For example, some natural biopolymers including lignin, cellulose, polyphenols, and proteins may persist for years in soil because of their polymeric structure, crystallinity, or chemical stabilization through interactions with other substances, such as protein-tannin complexes (Oades, 1988b). In most soils where biological processes are not unduly restricted, the amount of these polymeric substances is likely to be small, because even such a structurally complex molecule as lignin can then be decomposed.

In general, further research on the chemical constitution of nonhumic substances, whether in temperate or tropical soils, would be of questionable value. Some research may be warranted, however, to understand the specific pathways involved in organic matter turnover or to improve soil test procedures for plant nutrient availability from organic sources. An example is that of plant waxes in soil and their possible role in the formation of stable humic substances (Oades, 1988b).

Data on functional attributes—such as pK's, density of dissociable groups, and complex formation constants—would be of immense help, however, for modeling pedogenesis and soil acidification.

Humic Substances

General Characteristics

Humic substances are widely considered to represent the most stable fraction of SOM. Their stability is variously attributed to their chemical structure and heterogeneity as well as to entrapment within soil aggregates and interactions with metal cations and clay minerals. The recalcitrance of humic substances may owe more to the latter than to the former mechanisms since it has been shown that when extracted humic acids are incubated with fresh soil, they rapidly decompose unless polyvalent cations are present (Juste et al., 1975). This conclusion is also supported by recent spectroscopic investigations of cultivation effects on humic substances in a tropical Vertisol (Skjemstad et al., 1986).

As a major reservoir of soil carbon and nutrients, humic substances appear to be involved in the survival strategy of a small portion of soil microorganisms by providing a small but steady source of food. The production of resting structures or stages is a much more important survival strategy, however.

Where comparisons have been made between humic substances from arctic, temperate, and tropical soils, in terms of elemental and functional group compositions, no major differences have been observed; moreover, chemical degradation studies indicate that the humic and fulvic acids from these soils are structurally alike (Schnitzer, 1977). This similarity in composition and structure between humic substances from soils developed under such widely differing climates, however, needs to be critically assessed.

No comparisons have been made between humic substances from soils of similar mineralogy where temperature is the only soil-forming variable. The data of Schnitzer (1977) for tropical SOM refer only to two soils derived from volcanic ash. Systematic studies of SOM composition along well-defined sequences of climate, mineralogy, and vegetation are required. Moreover, humic and fulvic acids may not truly represent all organic matter in a particular soil because only 50% or less of the organic matter in temperate mineral soils is extractable with alkali (Kononova, 1975). Even less is likely to be extracted from some common tropical soils (e.g., Oxisols and Andisols) where humus may be strongly bound to iron or aluminum sesquioxides (Fox, 1980; Wada, 1980; Higashi, 1983). In many tropical soils, fulvic acids appear to make up the major component of humic substances and consequently may play a greater role in nutrient cycling, soil aggregation, and pedogenesis. Furthermore, chemical extraction of humic substances from soil may produce artifacts that would vary with soil type. This is certainly true when humic substances are chemically degraded (Hayes and Swift, 1978; K. R. Tate and Theng, 1980). Last, in most of these studies, the light fraction was not removed before chemical extraction was carried out.

By using carbon-13 nuclear magnetic resonance (^{13}C NMR) spectroscopy (e.g., Wilson, 1987) and pyrolysis/mass spectrometry (Py/MS) (e.g., Bracewell and Robertson, 1987) some of these problems have been avoided. The application of these powerful analytical techniques to studying humic substances and the information they provide are discussed below.

Classification Schemes

As already mentioned, the use of alkali-extractable organic fractions has limitations for making comparisons between soils developed under different climates (Schuppli and McKeague, 1984). Nevertheless, certain parameters, such as the humic to fulvic acid ratio and the distribution of these humus fractions throughout the soil depth, reflect the parent soil group and the extent to which humic substances interact with clays and sesquioxides (Kononova, 1975; Tsutsuki et al., 1988).

A system, devised by Kumada (1987), classifies humic acid into four types: Rp, P, B, and A, which can be related to soil properties, vegetational type, and land use. Rp-type humic acids represent an early stage in the humification process (e.g., rotting wood), whereas the P-, B-, and A-types represent stable forms in strongly acid, moderately acid, and weakly acid-to-alkaline soils. Soils developed under forest produce P-type humic acids with distinctive absorption bands at 615, 570, and 450 nm. A-type humic acids in Japan are usually found in Andisols with a thick black topsoil. The accumulation of organic matter in such soils has been ascribed to the formation of aluminum- and iron-humus complexes (Higashi, 1983). Infrared, ^{13}C NMR, and other spectroscopic analyses indicate that much of the humic acid carbon occurs in aromatic structures. Aromaticity decreases in the order A $>$ B $\simeq$ P $>$ Rp while alkyl-C decreases in the reverse order (see Figure 10). A-type humic acids are also associated with grassland vegetation, and their unusually high aromatic-C content may, in part, be caused by a history of stubble burning (Almendros et al., 1988).

Kumada's (1987) classification system, based on spectral characteristics, has been used to compare the humic acids extracted from 32 Andisols from New Zealand, Chile, and Ecuador (Figure 9). The humic acids from the New Zealand and Chilean soils and half of those from the soils of Ecuador are of the B- or P-type, whereas those from Japanese Andisols are of the A-type. Generally no differences are found that could be attributed to climate. Vegetational and cultural factors are probably responsible for the difference between the humic acid types in these soils and those in Japanese Andisols since the formation of Al/Fe-humus complexes is likely to be a common feature for all Andisols.

Humic acids extracted from a range of New Zealand soils formed on parent materials other than volcanic ash also show a wide range of types (Yamamoto et al., 1987, 1989). ^{13}C NMR spectroscopy further indicates that Rp- and P-type humic acids are rich in alkyl-C, aromatic-C is dominantly

Figure 9. Classification of humic acids extracted from volcanic ash soils of New Zealand, Chile, and Ecuador using Kumada's (1987) system. Solid lines denote boundaries between R_p, B, P, and A-type humic acids. Rectangles indicate regions of the diagram where Japanese (broken lines) and New Zealand (solid lines) humic acids are most frequently found.

$\log K = K_{400}/K_{600}$, where K is the absorption coefficient at 400 and 600 nm

$RF = K_{600} \times 1000/\text{ml}$ of 0.1N $KMnO_4$ consumed by 30 ml of a humic acid solution

(Redrawn from Shoji et al., 1987, with permission from the publisher.)

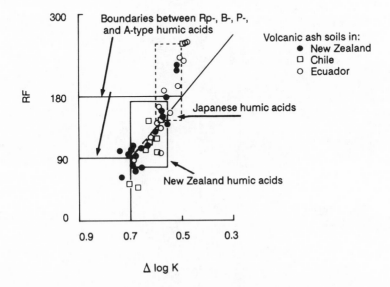

present in the A-types, whereas the B-types occupy an intermediate position (Figure 10). This approach may provide an insight into the composition of some of the more recalcitrant organic matter in soil. In most cases, however, the extracted humic acids represent a minor proportion of the soil organic matter.

Carbon-13 NMR spectroscopy has suggested that humic substances are generally not as aromatic as previously believed (Wilson, 1987). Indeed the relative prominence of alkyl-C in the ^{13}C NMR patterns of humic acids (see Figure 10) is a feature of the spectra. Caution is needed, however, in making quantitative assessments of composition from such spectra. This is because humic acids usually contain some "ash," which may include paramagnetic ions. Spectral broadening due to Fe^{3+}, for example, can result in some peaks being underestimated. This problem becomes especially serious for the humin fraction of humic substances and for whole soils. Nevertheless, the formation of humic acids during the decomposition and transformation of plant residues is often accompanied by increases in alkyl-C and carboxyl-C and decreases in ether-C, acetal-C, and aromatic-C (Zech et al., 1985).

Figure 10. Comparison of the carbon chemistry of a range of humic acid types from a Japanese and some New Zealand soils, obtained by cross polarization magic angle spinning (CP MAS) carbon-13 NMR spectroscopy. (Data of K.R. Tate, et al., 1990.)

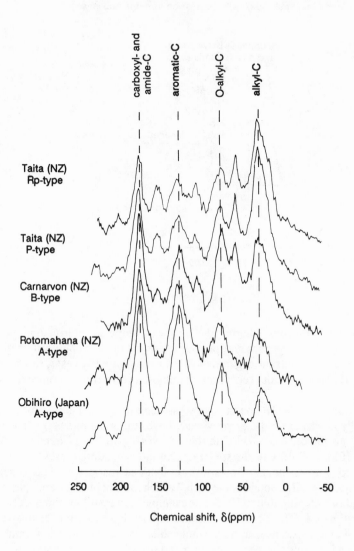

Dynamics of SOM in Tropical Ecosystems

Constitution and Stability

Besides being ubiquitous, alkyl-C, as in polymethylene structures, apparently represents a particularly recalcitrant form of soil C. In some tropical soils (e.g., Oxisols and Ultisols) in which organic matter contents tend to be low, alkyl-C may make up a relatively high proportion of SOM, as indicated by a recent ^{13}C NMR spectroscopic study of humic substances from a tropical Vertisol (Skjemstad et al., 1986).

We suggest that the constitution of organic matter in many tropical soils is dominated by alkyl-C. In support of this suggestion, Siem et al. (1977) and Lobartini and Tan (1988) found that the humic acids from tropical soils had a higher content of methyl and methylene groups and a lower content of carboxyl and aromatic groups than did soils from temperate latitudes.

Under the acidic conditions that often occur in tropical soils, the ionization of carboxyl groups would be suppressed. Polymethylene chains containing largely unionized carboxyls would then tend to adopt a random coil shape. This would enable the molecules to penetrate the micropores of clay domains, affording protection against microbial decomposition. Such a mechanism is also consistent with the observation (discussed later) that the fine clay fraction of soils is generally enriched in alkyl-C.

The climatic dependence of SOM constitution is also evident in a climosequence of New Zealand soils in tussock grassland (Molloy and Blakemore, 1974). In these soils, moisture regimes range from aridic to aquic and temperature regimes range from mesic to cryic. Here again, the organic matter composition does not differ qualitatively across the sequence. However, there are marked variations in the quantity of some constituents. Thus, polysaccharides are more abundant in soils at the cool, wet, high altitude sites than in soils developed under relatively mild, dry conditions (Bracewell et al., 1976; Molloy et al., 1977).

Investigation of 23 New Zealand soils, including those of the tussock grassland climosequence, by Curie-point Py/MS has also indicated that organic matter composition is quite similar for all soils (J. Haverkamp and K. R. Tate, unpublished results). In most instances, however, small but significant differences are observed among certain mass peaks that may partly be explained by climate, vegetational history, and clay mineralogy. For the climosequence soils, a multivariate statistical analysis of 20 mass peaks, ranked highest for discriminating between organic matter constitution, showed that covariant aromatic fragments versus carbohydrates give the best separation. Some of these aromatic fragments may be produced by cyclization of aliphatic polymers.

Bracewell and Robertson (1987c) showed that aliphatic (polymethylene) chains, heavily substituted with hydroxyl and carboxyl, occur in soils in which biological activity is high and organic matter turnover is rapid. In contrast, relatively unsubstituted aliphatic chains are characteristic of organic matter in soils in which activity is restricted, for example, by low temperatures or excess moisture. Thus, a homologous series of n-alkanes, n-alk-1-ene and α, ωalkadienes are the major pyrolysis products of a Typic Xerochrept from Spain (Saiz-Jimenez and de Leeuw, 1987).

These results suggest that highly aliphatic polymers feature prominently as constituents of organic matter in both temperate and tropical soils and that differences in their substitution patterns are apparently directly linked to the effect of climate and soil conditions on biological activity.

It would appear that in most soils the aliphatic constituent of organic matter is mainly associated with the clay fraction (e.g., Catroux and Schnitzer, 1987; Oades et al., 1987). In a South Australian Rhodoxeralf, for example, the alkyl-C represents about 56% of the carbon in the fine clay fraction (Figure 11). Since this alkyl-C resists solvent extraction, it is almost certainly attached to the clay, forming an organo-clay complex. In contrast, the O-alkyl-C, mainly carbohydrate-C, is concentrated in the largest particle-size fraction (250-1000 μm). Aromatic-C is more or less evenly distributed throughout all particle sizes, and is a relatively minor constituent (Oades et al., 1987).

Figure 11. Relative proportions of the various forms of carbon in particle-size fractions of a South Australian Alfisol (Rhodoxeralf), derived from solid-state carbon-13 NMR spectroscopy. Dashed line denotes the proportion of carbon as carbohydrates based on chemical analyses. (Redrawn from Oades et al., 1987, with permission from the publisher.)

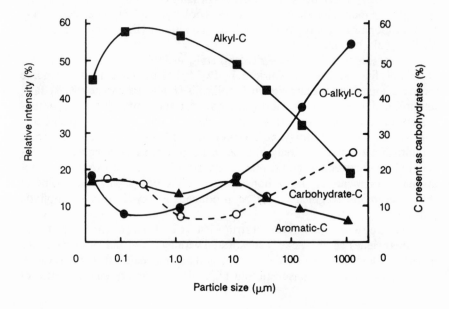

Dynamics of SOM in Tropical Ecosystems

Another consistent pattern is the decrease in C/N ratio with increasing particle density (see K. R. Tate and Theng, 1980; Sollins et al., 1984). The explanation for this observation should be sought in the increase in alkyl-C with decreasing particle size (see Figure 11). We propose that the centerbands with a chemical shift (δ) between 10 and 60 ppm in the ^{13}C NMR spectra of humic substances, assigned to aliphatic structures include a substantial contribution from amino acid side-chains (Newman et al., 1987). The exception is the signal at δ = 31 ppm, which is most likely attributable to polymethylene chain structures of lipids (see Figure 12) — that is, the organic matter associated with clay-size particles would be relatively enriched in nitrogen. Although this nitrogen is largely organic, inorganic forms, notably exchangeable ammonium, could contribute. This point needs to be checked.

Figure 12. Carbon-13 NMR spectra of air-dry clay fractions from a New Zealand Spodosol (8-16 cm depth). A was obtained by dispersing a sample of the air-dried soil in water and separating the clay fraction by gravity sedimentation. B is the clay fraction of the same soil previously treated with hydrogen peroxide followed by sodium dithionite-citrate-bicarbonate. (Reproduced from Theng et al., 1986, with permission from the publisher.)

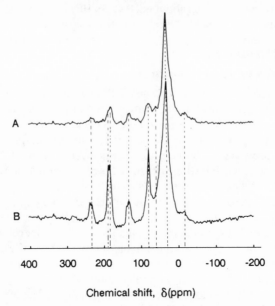

The persistence of alkyl-C structures with cultivation in a wide range of soil types and under different climates (Skjemstad et al., 1986; Preston et al. 1987) further suggests that these structures are highly recalcitrant. Whether this recalcitrance is an intrinsic property of the structure or is conferred by intimate association with clay mineral surfaces is still an open question. It seems, however, that both causes are responsible for the observed effect. First, the proportion of alkyl-C in SOM tends to increase as decomposition and humification progress (Zech et al., 1985). Second, polymethylene chains would adsorb on clay surfaces in an extended conformation, enabling strong van der Waals interactions to be established (Theng, 1979).

A highly aliphatic biopolymer, which is chemically akin to the aliphatic constituent of SOM, has been found in sediments (Nip et al., 1986). Because of its chemical stability and ability to survive over geological time, this polymer may act as an oil precursor. Similarly, polymaleic acids (Bracewell et al., 1980a) decompose very slowly when incubated with soil, although they can be strongly immobilized by clay minerals.

An excellent example of structural recalcitrance combined with physical protection by clay comes from the study by Theng et al. (1986) on two New Zealand soils from the climosequence referred to earlier. The organic matter in the clay fraction of these soils is dominated by polymethylene C (Figure 12), much of which is located in the interlayer space of the clay (a mica-beidellite). Being physically protected against microbial attack, this intercalated C is "inert." Accordingly, its age (about 6800 years BP) is much greater than that commonly found for topsoil C. Here a smectitic clay mineralogy, an accumulation of organic matter, and a highly acid soil reaction have combined to produce the "fossilized" organic matter.

Interactions with Inorganic Constituents

Because of its resistance to acid and alkali extraction, clay-associated organic matter may be classified as humin. Hatcher et al. (1986) have suggested that soil humins are not clay-humic acid complexes because their ^{13}C NMR spectra are quite different from those of humic acids. One reason for this difference could be the presence in humins of organic matter-iron precipitates that obscure NMR visibility. Paraffinic constituents of soil humins, which tend to be preferentially adsorbed by clay, could also lie behind the observed differences with humic acids. A third reason for the difference could be the presence of a light fraction in the humin.

Clearly the interaction of organic matter with clay minerals can influence the relative amounts of specific constituents without necessarily causing major qualitative differences in organic matter composition. Further investigation of the organic matter associated with clay minerals in similar soil types of temperate and tropical regions appears warranted before the effect of clay-organic interactions on organic matter constitution can be properly assessed.

Techniques other than ^{13}C NMR, however, may be necessary for quantitative comparisons between temperate and tropical soils. This is because the detection of carbon in whole soils and physical separates can be seriously affected by the presence of paramagnetic ions such as iron (Figure 13). Some improvement in spectral resolution can be achieved, particularly for O-alkyl-C and carbonyl-C, by pretreating the samples with sodium dithionite to remove iron (Oades et al., 1987; Wilson, 1987).

Figure 13. Effect of sample Fe content on CP MAS carbon-13 NMR spectra of some New Zealand soils and lignites before and after acid-washing with 5M HC1. NMR "visibility" is expressed as the carbon-13 signal per unit mass of carbon in the sample in relation to the signal for hexamethyl benzene per mass of carbon. The theoretical curve assumes a Poisson distribution of protons around the Fe nuclei in the sample, with a sphere radius of 1 nm. (Unpublished data of R.H. Newman, personal communication.)

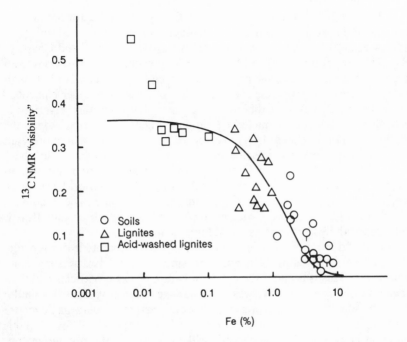

The organic matter associated with the fine clay fraction in some soils, notably those with high base status (e.g., Mollisols), has been shown to have a quite high turnover time (e.g., D. W. Anderson and Paul, 1984; Tiessen et al., 1984c). Some of this "young" carbon, however, could possibly become associated with the clay fraction during its physical isolation (by flotation) from soil. In new grassland soils, polysaccharides may represent a significant proportion of this clay-associated organic matter. In Mollisols the organic matter is mainly stabilized in 0.2-2.0 μm size aggregates. Cation bridging involving calcium and the functional groups in humic substances appears to be the principal mechanism of stabilization (Tsutsuki et al., 1988).

However, in leached soils, which are common to tropical regions, the most readily extractable organic matter tends to be concentrated in the 2-20 μm fraction (J. M. Oades, personal communication). Presumably, plant debris in this size range would be readily decomposed while the organic matter in the clay fraction (< 2 μm) is strongly linked to clays limiting its extractability. This organic matter is envisaged as fulvic acid-type polyanions, bound to oxide surfaces, plus some polysaccharides and N-compounds. Further work on humic substances in particle-size separates of different soils is needed before general conclusions about the quality, distribution, and stability of organic matter can be drawn.

The stabilization of organic matter through complex formation with aluminum and iron is common to many strongly weathered tropical soils (e.g., Andreux et al., 1985) and to Andisols (Higashi, 1983; Mizota and Chapelle, 1988). The propensity for Al and Fe to hydrolyze results in a range of positively charged, partly hydrolyzed, and polymerized species that would adsorb strongly on clay surfaces. Investigating the interaction process even under controlled pH conditions is therefore difficult. We know, however, that a variety of mechanisms are involved, including cation-bridging, anion exchange, and specific adsorption (Theng, 1979). As regards aluminum, its solubility means that Al plays an active role in soil acidity and in transport processes involving soluble organic matter. In acidic tropical soils containing oxic materials, organic matter strongly interacts with the oxide surfaces by cation-bridging and/or ligand exchange (Theng, 1987). The organic matter is likely to be dominated by fulvic acid-like substances with an aliphatic chain structure and strongly substituted with hydroxyl and carboxyl functional groups.

Use of density or particle-size fractionation to separate hydrous oxide-organic complexes is apparently not promising (J. M. Oades, personal communication). This is because the colloidal nature of oxide particles and their association with organic matter results in a range of effective densities similar to those of other soil colloids present. However, natural oxide-organic matter complexes have been successfully separated from tropical soils by magnetic separation (Hughes, 1982) apparently without oxidizing the organic matter. Used in combination with a technique like Py/MS, for example, it may be feasible to determine the constitution of the organic matter in such complexes.

Constitution Using Stable Carbon Isotopes

Carbon inputs and turnover rates can be estimated from changes in the stable C isotope composition of soil organic matter. This approach makes use of the characteristic $^{12}C/^{13}C$ isotope ratio of the vegetation as a marker to follow the dynamics of soil C, which largely derives from plant residues. Dynamic processes usually cause only comparatively small isotopic fractionation, although this can be a problem in some soils (Volkoff and Cerri, 1987).

As a result of differences in their photosynthetic uptake of CO_2, plants are more or less enriched in ^{13}C. For example, C_4 plants, which include tropical Gramineae (e.g., sugarcane and maize), assimilate more ^{13}C than C_3 plants to which most temperate plants belong. By growing C_4 plants in soils previously under C_3 plants, we can trace the fate of the C_4 plant carbon as it enters the soil and is either incorporated into the soil organic cycle or respired. This enables the size of the labile and stable pools of C and the C turnover rates to be measured by what is essentially an *in situ* labeling technique (Cerri et al., 1985; Balesdent et al., 1988).

Figure 14 shows the data of Balesdent et al. (1988). In this instance, the carbon from the original native prairie still comprises 61 and 49%, respectively, of the total soil C, even after 100 years under grass and wheat. The higher level of organic C under grass was ascribed to the presence of more labile, physically protected C than the C in the soil under wheat in which protection is lost through annual tillage operations. The initial rapid decline of the prairie organic matter in the first 27 years of cultivation graphically illustrates the loss of the most labile organic matter mainly through rapid mineralization by soil microorganisms.

Figure 14. Changes in the amount and origin of soil organic C accompanying long-term cultivation of wheat on formerly virgin prairie soil. Circles are the 0 to 10 cm depth sample and lower circles are the 10 to 20 cm depth samples. (Redrawn from Balesdent et al., 1988, with permission from the publisher.)

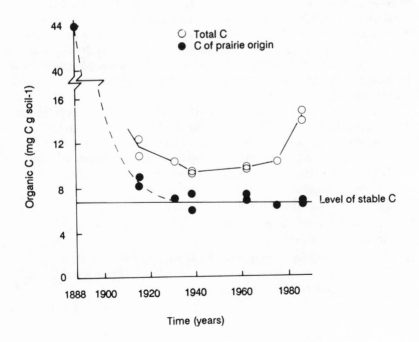

Using a similar approach, Cerri et al. (1985) have also distinguished between labile and stable C for a tropical soil in which sugarcane was grown after having been under forest.

Partitioning of the soil organic C between prairie grasses, C_3 forage crops, and maize enabled Balesdent et al. (1988) to contrast the stability of the carbon from the original vegetation with the carbon added over the past 61 years. Particle size and density separations of this soil indicated that the fine clays were mainly responsible for stabilizing organic matter, although they also contained some labile C. Evidence was also provided for the presence in larger particle-size fractions of organic matter of more uniform stability, presumably resulting from further protection within aggregates (e.g., K. R. Tate and Churchman, 1978; Tiessen and Stewart, 1983; Balesdent et al., 1987).

This general pattern of increasing organic matter stability with decreasing particle size is consistent with previous reports (e.g., Tiessen et al., 1984c). A major advantage of the natural ^{13}C-labeling technique for soil organic matter studies over that of ^{14}C-labeling is that labeling has occurred *in situ* over thousands of years so that even the most stable fractions are included.

Similar soil types should be used when comparing temperate and tropical soils. For example, in two New Zealand grassland soils with quite different mineralogies, organic matter declined under continuous maize (Cotching et al., 1979). In the Typic Dystrandept, however, this trend stabilized after 6 years, presumably because the allophane here confers great stability to the organic matter (Zunino et al., 1982; Legay and Schaefer, 1984). This is reflected in the different patterns of ^{13}C enrichment under maize (K. R. Tate and G. L. Lyon, unpublished results). Most of the organic matter remaining in the Typic Dystrandept (when levels have stabilized) is not involved in a rapid turnover. In contrast, for the Aquic Dystrochrept where halloysite is the dominant clay mineral, isotopic composition of the organic matter gives no indication of a significant stable organic matter fraction. That is, organic matter levels continue to decline.

The wide distribution of Andisols in both temperate and tropical regions of the world (K. R. Tate and Theng, 1980) would make them good candidates for comparative research on organic matter constitution and dynamics.

Use of stable C isotope composition of SOM as a means of comparing the dynamics of C incorporation and turnover in and between temperate and tropical soils is suggested as a research imperative.
This approach offers the following advantages:

1. Plants with C_3 and C_4 photosynthetic pathways are found both in temperate and tropical regions.

2. The *in situ* labeling technique can be applied across ecosystem types within tropical and temperate regions (e.g., tropical rain forest through tropical Gramineae, in all soil types).

3. Completed studies within each region have demonstrated the feasibility of using stable C isotope composition as an index of organic composition for dynamic studies.

4. Whole soils and physically (rather than chemically) separated fractions are used, thus minimizing artifact production during fractionation.

Care may be needed, however, because some isotopic fractionation occurs between different plant components (Benner et al., 1987) and during pedological processes (e.g., Volkoff and Cerri, 1987).

Conclusions and Recommendations

Constituents of SOM in tropical soils are qualitatively similar to those found in temperate ecosystems. The observable variation in quantity and composition may result more from past climatic and edaphic conditions than from current conditions. Soil clay content, pH, and mineralogy are known to have major effects on SOM properties. The emphasis of future research should therefore be directed toward understanding the causes behind this variability of SOM constituents within a climatic region and along defined soil gradients. Because of the constitutional similarity between tropical and temperate SOM, many of the methodologies developed for temperate soils may be adapted for tropical SOM studies. These procedures must provide for analyses of both the living and dead SOM constituents. Analyses of these constituents must take into account their relative stability in soil; thus, their quantity and turnover rates need to be measured. Accordingly, we make the following recommendations for research on the constituents of SOM in tropical ecosystems.

Theme Imperative: The amount, composition, and dynamics of living and dead organic constituents need to be assessed using similar soil types within a climatic region.

Research Imperatives:
1. Assess the quality of SOM in whole soil and biologically meaningful soil fractions along well-defined gradients of climate, soil chemistry, and mineralogy.

2. Determine the effect of soil factors and resource quality on the decomposition of, and nutrient release from, litter. This should also extend to changes that occur in constitution as decomposition proceeds.

3. Organic matter transformations in soil result from the interaction of fauna and microbial populations. Elucidate and quantify these interactions.

4. Quantify labile SOM constituents in tropical soils, especially the light fraction, available C, and microorganisms. Assess their impact on ecosystem function.

5. Critically assess the assertion that SOM constituents are stabilized mainly by their interaction with mineral constituents and quantify the stability of the organo-mineral complexes and their aggregates in different soil types.
6. Quantify the impact of land use and management on SOM constituents.
7. Develop fractionation schemes that truly reflect the dynamic character of SOM for use on a wide range of soil types.
8. Collate and assess local reports and unpublished data on SOM constituents from laboratories in developing countries and integrate them into a common data base for modeling purposes.

Acknowledgments. We thank C. M. Preston, M. van Noordwijk, and K. Wada for helpful comments and suggestions.

Dynamics of SOM in Tropical Ecosystems

Chapter 2

Soil Organic Matter as a Source and a Sink of Plant Nutrients

John M. Duxbury, M. Scott Smith, and John W. Doran
with Carl Jordan, Larry Szott, and Eric Vance

Abstract

Nutrients in soil organic matter (SOM) are present within complex polymers and mineral-organo associations. Investigations into the chemical composition of SOM and the nutrient elements within it have only been partially successful and have not been linked to biological processes. Particle-size and density fractionation studies have provided insight into the architecture of soils and on the distribution and stability of SOM and nutrient elements. Such studies should be continued.

Modern concepts of mineralization and immobilization of N, P, and S in soils recognize that SOM is heterogenous with respect to biological activity, that mineralization-immobilization processes occur simultaneously, and that microbial biomass itself represents a significant sink or source of nutrients. The stability of SOM largely results from protective processes occurring within soils rather than from the creation of recalcitrant chemical structures. Although the stability of SOM is likely to be continuously variable, specific pools are recognized for mathematical modeling purposes. These commonly include microbial biomass (BIO), an unprotected or labile pool (LAB), a pool protected only in the absence of cultivation (POM), and a pool that persists for very long time periods

(COM). At present, the more stable pools are estimated from simulation modeling and cannot be directly measured. Isotopic studies have great potential to help this situation. Questions remain about the validity and utility of both the chloroform fumigation procedures — which are widely used to determine microbial biomass C, N, P, and S — and laboratory N mineralization procedures. In addition, a better understanding is needed of the effects of multiple interacting environmental variables on both microbial dynamics and microbially mediated processes. Most of the effort in understanding the behavior of labile organic matter has been directed to C and N and more attention should be given to factors controlling P and S cycling, which may be partially independent of C and N.

Conversion of land to agricultural use makes a fraction of the protected SOM susceptible to mineralization, disrupts internal recycling of nutrients, and increases the potential for loss of nutrients from the system. Opportunities exist to develop soil and crop management systems that result in better synchrony between nutrient mineralization and plant uptake and conservation of nutrients not used by crop plants. Tillage management, which can be an important tool for the manipulation of nutrient storage and release, has not been investigated to a large extent in the tropical environment.

Soil organic matter is often an important source of negative charge in tropical soils and as such promotes retention of available forms of base cations throughout the soil depth. Many of the negatively charged sites in organic matter, however, are blocked by interactions with Fe, Al, and oxide surfaces. The source of "organic" CEC and its maintenance needs to be identified in terms of the pools of organic matter previously described.

The long-standing issue of whether or not crop yield potential is affected by SOM cannot be resolved solely in terms of nutrient availability. It is important to separate effects due to organic matter per se from those due to its decomposition.

Significant advances in both understanding and managing the behavior of soil organic matter (SOM) as a source or a sink of plant nutrients will only be achieved through studies at a conceptual level. Consequently, this paper considers the mechanisms that regulate nutrient availability rather than empirical or statistical relationships, which abound in both the soil fertility-crop production literature and the ecological literature. Although we recognize the great diversity in tropical soils, we have chosen not to discuss individual soil orders because basic principles apply to all soil types and because we have a poor understanding of how a variety of incompletely characterized biological, chemical, and physical processes interact in many different environments to determine SOM behavior. We have tended to make comparisons between temperate- and tropical-region soils and to discuss agroecosystems because most of the recent advances in our understanding of the composition and behavior of SOM have come from studies in temperate-region agroecosystems. At the same time, we realize that information from temperate-region soils cannot always be extended to tropical soils because differences in soil chemistry and mineralogy strongly influence the behavior of SOM.

It has long been accepted that one of the most important, and certainly the most studied, contributions of SOM to soil fertility is its capacity to supply nutrients for plant growth, especially N. Nutrients are sequestered in, or released from, SOM by two fundamentally different processes: *biological processes* control storage and release of N, P, and S as these elements are contained in structural units of SOM, whereas *chemical processes* control interactions with macro- and micronutrient cations (Ca, Mg, K, Fe, Cu, Zn, and Mn). To some degree these two processes create a paradoxical situation in that mineralization of SOM, which is required to release N, P, and S, reduces the capacity of soils to interact with cations that would be available to plants through chemical equilibration processes. Because SOM is often the major source of negative charge on tropical soils, its maintenance is important for retention of available forms of cations within the soil and must be balanced against any desire to exploit organic N, P, and S reserves.

N, P, and S in Soil Organic Matter
Amounts and Distribution

The accumulation and distribution of SOM and its associated nutrients within the soil primarily depend on the quantity and distribution of organic residue inputs (largely plant residues), on the rates of biological decomposition processes, and on the capacity of the soil to protect SOM (mostly humic substances) from microbial decomposition.

There has been some disagreement about the relationship between latitude and total SOM and soil nutrient storage. Variation in total soil C within the tropics is clearly greater than differences across latitudes. Thus relationships between temperature or latitude and total SOM are difficult to establish unless the influences of mineralogy, management, moisture regime, and other factors are also considered. Sanchez et al. (1982) found no effect of present climate on mean levels of SOM in given soil orders when comparisons were made of agricultural sites. In a study of over 3000 soil profiles from natural ecosystems, Post et al. (1982, 1985) examined total soil C and N as a function of latitude, temperature, and moisture regime. Although their data suggest that soil C and N decrease with decreasing latitude, moisture regime was found to be a more powerful determinant than latitude or temperature. For climatic zones with similar moisture regimes—for example, moist forests—total soil C and N were not consistently correlated with temperature or latitude.

Post et al. (1985) further noted that the C/N ratio was related to temperature, being generally, but not consistently, lower in warmer climates. This results from a more rapid and complete decomposition of labile organic matter. Useful comparisons across climates were possible in this study because all samples came from areas with natural vegetation; no large, comparably homogeneous data base exists for cultivated soils.

Within the tropical environment, the accumulation and distribution of SOM throughout the soil depth is generally affected by ecosystem type. Because inputs of residues to surface soil are higher in tropical forests than in savannas and savannas are more subject to residue loss by fire, more SOM is found in soils under tropical forests than in those under savanna vegetation, with much of the difference being in the top 10 cm of soil.

Generally, 95% or more of the N and S and between 20 and 75% of the P in surface soils are found in SOM. Mean values for the organic C/total N/total S ratio of surface soils are fairly constant across the range of soil types and climates, but generalized differences have been noted between cultivated and virgin soils. Thus, the mean C/N/S ratio of agricultural soils is about 130:10:1.3, whereas that of virgin soils, both grassland and forestland, is about 200:10:1 (Freney, 1986; Stevenson, 1986). These differences can be caused by one or more of the following: (a) preferential mineralization of C relative to N and S, and of N relative to S in cultivated soils, (b) the generally higher nutrient concentration of agricultural crop residues, and (c) differences in retention of the various elements in the soil-plant system after mineralization. The mean value for P is close to that for S, but the P content of SOM is more variable than the other elements. In fact, the C/N, C/P, and N/P ratios can vary widely as a function of parent material, degree of weathering, vegetation, and management, whereas the N/S ratio is less variable and usually within the range of 6-8:1 (Biederbeck, 1978; Freney, 1986). The C/P and N/P ratios of organic matter in tropical soils seem to be especially variable; some are considerably wider (say by a factor of 2) than those in temperate-region soils (e.g., Nye and Bertheux, 1957; Enwezor and Moore, 1966), but others are well within the range found in temperate-region soils (e.g., Bornemisza and Igue, 1967; Islam and Ahmed, 1973; M. J. Jones and Wild, 1975; Neptune et al., 1975; S. J. Smith and Young, 1975; Sharpley and Smith, 1983; Mueller-Harvey et al., 1985). The greater constancy of soil N/S ratios is interpreted to indicate coupling of their cycling in the soil-plant system (Barrow, 1961; Biederbeck, 1978) and the high variability of P to a degree of independence in the cycling of soil organic P relative to C, N, and S. Possible reasons for variability in soil C/P values and for differences in element cycling are discussed later in this chapter.

It has been suggested that highly weathered soils of the tropics (Oxisols and Ultisols) usually contain less total P and have a higher proportion of organic

Dynamics of SOM in Tropical Ecosystems

P than do the young, less-weathered soils of temperate regions. R. A. Olson and Englestad (1972) gave average total P values of 200 mg P kg^{-1} soil for highly weathered tropical soils and 3000 mg P kg^{-1} soil for soils of subhumid temperate regions. The literature supports the 200 mg kg^{-1} figure for tropical soils, but the figure for temperate-region soils seems to be too high; for example, reported values for U.S. soils average about 1100 mg P kg^{-1} (Grove, 1985). Nevertheless, with some exceptions, the trend suggested by Olson and Englestad appears to hold.

Sanchez (1976) considers that 60-80% of the P in tropical soils is usually organic compared to 20-50% for temperate-region soils, but data from the West African savanna (Enwezor and Moore, 1966; M. J. Jones and Wild, 1975; Mueller-Harvey et al., 1985), Brazil (Neptune et al., 1975), and Costa Rica (Bornemisza and Igue, 1967) mostly fall within the range given for temperate-region soils. Whether or not tropical soils contain a greater proportion of their P in organic forms, their reduced total P content and sometimes high phosphate adsorption capacity accentuates the importance of organic P as a source of P for plant growth in these soils.

Forms

Soil chemists have spent a considerable amount of time trying to identify the N-, P-, and S-containing organic structures in soils. The nutrients are almost totally present in complex polymers in associations with other organic and soil mineral components, and isolation of them from whole soil is difficult. The general approach used in these studies has been to subject soil or extracted organic fractions to various separation and chemical procedures, such as hydrolysis, to liberate constituent monomers, which can then be identified individually or by compound class. The present state of this methodology is far from perfect, and considerable loss or alteration of many individual compounds undoubtedly occurs. Research to date is as follows.

Organic N

The work on soil N has been well reviewed by Stevenson (1982b, 1986). In general, only about 40-50% of the organic N in soils can be positively identified as belonging to particular chemical classes. Amino acids and amino sugars are the only quantitatively significant components identified to date. Reported ratios of amino acid N to amino sugar N range from 1 to 46:1, although amino acids usually dominate. Variations in methodology and methodological problems make absolute values and comparisons among different studies highly questionable.

One of the more interesting observations of the chemical work with soil N has been that while newly immobilized ^{15}N has a different fractionation pattern than native soil N (more amino N and hydrolyzable unknown N and less nonhyrolyzable N), it quickly becomes similar to that of the native soil humus. Although interpretation of this observation is subject to the limitations of the fractionation scheme, it suggests that newly added N is rapidly converted to the same chemical structures present in old soil N. If this is correct, the more rapid mineralization of recently added N must be due to differences in substrate availability or protection rather than to differences in structural composition of new and old N. This conclusion is supported by results from long-term field experiments that show little impact of either tillage-induced declines in SOM levels or different cropping systems on the distribution of N forms.

Organic P

Most of the soil organic P is present as esters of orthophosphoric acid, and numerous monoesters and diesters have been isolated from soils (Halstead and McKercher, 1975; G. Anderson, 1980). Small amounts of phosphonates, which contain P bonded to C rather than to O, have been detected in extracts of soil by NMR (nuclear magnetic resonance) (K. R. Tate and Newman, 1982; Hawkes et al., 1984). NMR measurements also show that most of the 30-80% of the organic P that can be extracted from soil is present as monoesters. Phosphate esters of inositols, the cyclohexane analogs of hexose sugars, are the most abundant identifiable compound class, ranging from 5 to 80% of the soil organic P. The hexa- and pentaphosphate forms are the most common in soils. The accumulation of inositol phosphates in a wide variety of soil types is thought to be due to their capacity to form insoluble precipitates with Fe, Al, and Ca and to strongly adsorb on amorphous Fe and Al oxide surfaces (G. Anderson and Arlidge, 1962). In acid soils from Scotland inositol hexaphosphate occupied the same adsorption sites and was more strongly adsorbed than orthophosphate. Inositol-P adsorption was correlated with oxalate-extractable Fe + Al, but adsorption of penta- and hexa-inositol phosphates naturally occurring in soils was best correlated with Fe alone. The quantity of inositol phosphates in Canadian and Scottish soils was correlated with orthophosphate adsorption capacity (McKercher and Anderson, 1968; G. Anderson et al., 1974). A similar result was obtained with 16 acid soils from Bangladesh, where inositol phosphates accounted for 10-82% of the organic P (Islam and Ahmed, 1978). Mineralization of inositol P in two of these soils with high inositol P content was slow unless soil pH was raised by liming or submergence (Islam and Ahmed, 1973).

Other identified components of soil organic P include nucleotides and phospholipids, but both groups are considered to be minor contributors, with nucleotides accounting for less than 5% and phospholipids accounting for less than 15% of the total. Since both groups are components of all living tissues, they may, to some extent, be being extracted from soil organisms and plant roots. Except in soils that contain large amounts of inositol phosphates, structures of phosphate esters in soil are largely unknown.

Dynamics of SOM in Tropical Ecosystems

Organic S

The principal form of sulfur added to soils is amino acid S, mostly composed of methionine, cysteine, and cystine, which are also commonly isolated from acid hydrolyzates of soil and extracted SOM. Sulfur-containing amino acids account for up to 30% of the organic S (Freney, 1986). Between 30 and 70% of the organic S in soils can be reduced to H_2S by HI. Most of the HI-reducible S is thought to be ester sulfate (C-O-S), although C-N-S and C-S-S (as in cystine) bonded sulfur would also be included; carbon-bonded S is not reduced by HI.

Other Structures

From the foregoing summary of our knowledge of the organic forms of N, P, and S in soils, it is evident that at a monomer unit level these elements are found to some extent in structures that are similar, if not identical, to those in which they are added to soil. Exhaustive chemical studies using the best methodology, however, usually identify fewer than half of the specific organic chemical forms of any of these elements in soils. This has led to considerable speculation about the presence of other types of structures, many of which could be formed by chemical reactions involving a wide variety of plant- and microbially derived products. Most of the speculation is over forms of N, e.g., heterocyclic N, including pyridine and phenoxazone derivatives (Flaig, 1975), and while many of these structures would be plausible components of soil humic substances, there is as yet scant evidence that any of them are actually present in SOM.

Comparisons of elemental ratios and structural forms of nutrients present in SOM have not shown any substantial differences between tropical and temperate-region soils. These comparisons do show, however, that such parameters vary widely among soils. Explanations for this variability are lacking.

Organic C, N, S, and P and Aggregate Size

Evaluation of the content and susceptibility to mineralization of organic C, N, S, and P as a function of aggregate size has provided another approach to studying the behavior of SOM. Most studies show that organic C, N, P, and S contents increase and C/nutrient ratios narrow with decreasing aggregate size (Chichester, 1969; Cameron and Posner, 1979; D. W. Anderson et al., 1981; Hinds and Lowe, 1980). In contrast, Elliott (1986) found that the C, N, and P content of aggregates decreased with aggregate size; however, C/N and C/P ratios narrowed with decreasing aggregate size. Some differences in the distribution of the various nutrients have been noted: D. W. Anderson et al. (1981) found that S was preferentially associated with fine clays and Elliott (1986) found greater P accumulation relative to C and N in fine fractions.

Most studies on mineralization of organic matter and nutrients in different sized aggregates (Chichester, 1969; Cameron and Posner, 1979; Lowe and Hinds, 1983) show that both organic matter and nutrient mineralization increase with decreasing aggregate size. However, Elliot (1986) found greater mineralization of C and N in macroaggregates ($> 300 \mu$m) than in microaggregates (50-300 μm), and postulated that mineralization of interaggregate organic matter is the main source of nutrient release as SOM levels decline with cultivation. There is also evidence from comparison of cultivated and uncultivated soils that organic matter and nutrients in the fine silt/coarse clay fractions are the most stable. Thus, Tiessen et al. (1983) found that cultivation had the least effect on organic P content in the fine silt coarse clay fraction and Tiessen et al. (1984a) found that the natural ^{15}N abundance of N in this fraction was unchanged upon cultivation. In contrast, organic matter and nutrients associated with fine clays are readily mineralizable. Studies by Ladd et al. (1977a, 1977b) on immobilization and remineralization of ^{15}N in soil also suggest that over time N will accumulate in the coarse clay/fine silt fraction.

Research Needs

In vitro studies of the amounts and forms of nutrients in SOM have little relevance to the behavior of SOM as a source or a sink of nutrients in soils. There is still a pressing need to link chemical and biological studies; chemists tend to fractionate and study components of SOM without regard to the biological significance of the fractions, while biologists and ecologists have developed concepts and models without fully considering the chemical or physical logic of pools of differing biological stability or what can be exerimentally measured.

More information is needed on the chemical forms of organic phosphorus (and perhaps also sulfur) in tropical soils and on their interactions with soil mineral components. Most attention should be given to soils with high phosphate adsorption capacity; these most likely will be acid soils in which strong interactions with active forms of Fe and Al occur. Chemical studies should be extended to determine the bioavailability of organic P and S.

Particle-size and density fractionation procedures provide information at a more detailed level than do studies with whole soil and are beginning to probe the architecture of SOM in soils. Studies along these lines should be continued within a wide range of tropical soils, especially the Oxisols and Ultisols that have exceptional natural soil structure. At the same time, a word of caution: it is important that thought be given to questions of methodology. For example, is air-drying of soils — used by some researchers — appropriate, or does it affect both organic-matter distribution and mineralization? Similar questions should be asked about ultrasonic dispersion.

Biological Mineralization and Immobilization of N, P, and S

Concepts and Models

Gradually over the last few decades, a new and in some ways increasingly complex concept of mineralization and immobilization of nutrients in SOM has evolved. This model attempts to account for, and is in part based upon, the realization that:

1. Microbial biomass constitutes a significant sink and source for nutrients.
2. The process of decomposition is also a process of microbial synthesis.
3. Mineralization and immobilization occur simultaneously in soils.
4. Soil organic matter and the nutrients within it are heterogeneous with regard to biological activity; a fraction cycles very quickly, but some components cycle very slowly.

The process of nutrient mineralization and immobilization represents the net result of several interacting subprocesses, involves many functionally diverse organisms, acts on heterogeneous substrates simultaneously, includes multiple interacting elements, and is subject to multiple interacting environmental controls. This makes it an interesting, worthwhile, and frequent target for simulation models and much of our discussion will be framed in terms common to most of these models.

Environmental Regulation of Mineralization and Immobilization

The important environmental determinants of nutrient turnover and storage via SOM include soil chemistry and mineralogy, soil and vegetation management, and climate.

The effects of climate on biological processes are central to any discussion of the behavior of organic matter in soils. Fairly constant warm temperature is the defining climatic characteristic of tropical soils. More than adequate experimental information exists on the responses of microbial populations and processes to temperature change. Over the temperature range of interest, rates of individual processes increase rather predictably with increasing temperature (Q10 = approximately 2). Therefore it is expected that microbially-mediated processes, including turnover of nutrients in plant and animal residues, microbial biomass, and SOM, would generally be accelerated in tropical soils.

Integrating the effects of temperature change on multiple interacting processes and predicting interactions with other environmental parameters in soils is not so simple, however. Attempts to relate overall SOM levels to temperature or to climatic variables in general are a good example. They meet with mixed success because other variables also influence the level of stabilized organic matter in soils, which usually accounts for the bulk of the SOM. Climatic variables are good predictors of residue decomposition processes and of the turnover rate of unprotected organic C, N, P, and S (see later in this chapter).

Effects of Seasonality and Environmental Variation on Processes of Decomposition and Accumulation

It is widely recognized that seasonality and short-term environmental variation are critically important for controlling soil processes. Yet the mechanisms and magnitude of these effects have not been well defined, especially under field conditions. Temperature variation is less in the tropical environment and freeze-thaw effects are unimportant; however, soil wetting and drying cycles may be more extreme in some tropical regions. Fluctuations in both temperature and moisture are known to accelerate organic matter decomposition and N mineralization, although wetting and drying has probably received the most attention (Birch, 1958; Ladd et al., 1977a, 1977b). It is commonly stated that these effects are due to disruption of physically protected organic matter complexes and to accelerated turnover of microbial biomass, but this has not been clearly demonstrated. The short-term accelerating effects of temperature and moisture fluctuations may be quite different from the long-term effects of seasonal variability. It has been suggested that strongly seasonal environments favor accumulation of soil organic matter (for example, Harmsen, 1951). The most plausible explanation is that strong seasonality tends to inhibit decomposition more than primary productivity. Jordan (1985) concludes that as temperature increases, dry seasons will inhibit primary productivity more than decomposition; thus, seasonally dry tropical areas should have less SOM than strongly seasonal temperate zones where subzero temperatures inhibit both producers and decomposers. The data of Post et al. (1982) provide some evidence for this. Some models have considered the effects of environmental fluctuations (van Veen et al., 1984), but it is not certain how hysteretic biological responses should be quantitatively described. This is a difficult but unavoidable problem in describing quantitative relationships between climate and soil processes.

Nutrient Pools and Nutrient Availability

Defining the quality, availability, and activity of organic nutrient substrates and reservoirs is one key to understanding and describing mineralization-immobilization processes. This is a considerably more difficult problem for nutrients in SOM than for nutrients in plant litter. SOM in different soils differs greatly in quality or ability to supply nutrients. It has been apparent, at least since the classic study by Jansson (1958), that SOM cannot be considered homogeneous with regard to potential for mineralization. Isotopic techniques demonstrate that a fraction of the SOM and immobilized nutrients can be remineralized very rapidly, while ^{14}C dating shows that some soil organic C and presumably also the nutrients associated with it have a turnover time of the order of 1000 or more years.

Most current models describe SOM quality and nutrient availability in terms of discrete organic matter fractions, or pools, which vary in activity toward decomposition. Conceptually useful models will include at least pools for microbial biomass, nonliving but available or active C and nutrients, and stabilized or recalcitrant forms of C and nutrients.

Many of the models of SOM dynamics have evolved from the five-pool model of Jenkinson and Rayner (1977), which divided plant residues into decomposable (DPM) and resistant (RPM) fractions and included microbial biomass (BIO) and two forms of stabilized organic matter, named physically protected (POM) and chemically protected (COM) organic matter. Other models (McGill et al., 1981; van Veen and Paul, 1981; Molina et al., 1983; Parton et al., 1983; van Veen et al., 1984; van Fassen and Smilde, 1985) are more complex and use somewhat different terminology, but all have the same general structure. Jenkinson and Raynor's POM pool, and its equivalent in other models is included in models because of the long-term effects of cultivation on SOM levels; i.e., POM is temporarily protected under conditions of no-tillage. For the most stable, oldest COM fraction the mechanism of stabilization has not been clearly defined.

The partitioning of SOM into discrete pools also has some practical mathematical advantages. Complex and difficult kinetics in observed data can usually be adequately described as the sum of multiple components with simple kinetics, generally using first-order reactions. The major limitation to the discrete pool approach is the obvious one, i.e., that the system is likely to be continuously variable. Given this, the concept of pools and the models based on them will never completely match with biological, chemical, or physical reality.

Furthermore, it is not at all clear how these fractions are to be experimentally measured. There can be no completely satisfactory chemical or physical analysis for active nutrients, since availability is conditional, not inherent to the chemical or physical forms of the nutrients – i.e., it depends on interactions with other components and environmental conditions. The size of the active fraction of any nutrient does not define an absolute amount of a specific class of molecules but rather provides a relative index of biological nutrient availability under a particular set of conditions. Experimentally, it seems that this must be approached by biological incubation, perhaps combined with controlled physical manipulation. The most straightforward and widely used measurement of an active nutrient fraction is the Stanford and Smith (1972) aerobic N-mineralization procedure, in which rates of inorganic N accumulation under defined conditions are used to estimate the size of a potentially mineralizable N pool. N mineralization under laboratory conditions was described by a first-order kinetic model, and it was proposed that N-mineralization rates in the field could be calculated by using field temperature and moisture measurements to modify the decomposition rate constant (Stanford et al., 1973; Stanford and Epstein, 1974).

Subsequent studies (Verstraete and Voets, 1976; Cabrera and Kissel, 1988b) and calculations using field temperature and moisture data (J. M. Duxbury, unpublished data) have shown that the method can substantially overpredict N mineralization in the field. Various authors have criticized Stanford and Smith's mathematical methods and choice of a single pool model but have not field tested their proposed alternatives (e.g. Tabatabai and Al-Khafaji, 1980; Naske and Richter, 1981; Talpaz et al., 1981; Juma et al., 1984; Deans et al., 1986). However, Bonde and Lindberg (1988) demonstrated that estimates of N_0 (the soil's N mineralization potential) varied considerably as a function of the regression models employed to describe the N release pattern. Length of incubation (Paustian and Bonde, 1987) and soil pretreatment (Bonde and Lindberg, 1988) are additional factors that affect estimates of both N_0 and rate constants. Beauchamp et al. (1986) and Cabrera and Kissel (1988a) also showed that soil pretreatment, principally air-drying, can lead to higher decomposition rate constants or to a larger potentially mineralizeable N pool than in soils that are not dried. The widespread use of air-drying is unfortunate because it likely invalidates the results of a large number of studies and obscures analysis of the value of this approach. For example, possible valid explanations for overprediction of N mineralization include reimmobilization of mineralized N when plants are present in soils, a need for very detailed input of temperature and moisture data, e.g., hourly observations of temperature at small depth increments, and interactions between fluctuating moisture and temperature regimes not included in the procedures for modification of the rate constant.

A more complex but complete measurement of N activity was provided by Paul and Juma (1981). They fractionated organic N into microbial biomass (24-week relative half life), active nonmicrobial biomass (77-week half life), stabilized (27-year half life), and old (600-year half life). Biomass N was estimated directly by chloroform fumigation. Total active N was determined from total N and ^{15}N mineralized during laboratory incubation with the assumption that recently immobilized ^{15}N had uniformly mixed with the active N pool but had entered no other fraction. An old C fraction was estimated from ^{14}C dating and a C/N ratio for this fraction was assumed. Stabilized N was what remained. Although several of the assumptions used in these calculations are questionable, this remains one of the few attempts to assign values to experimentally and conceptually meaningful organic N fractions.

Support for the existence of a young, labile pool of SOM, behaving independently of older SOM pools and not interacting with other soil components, has come from elegant field studies using ^{14}C labeled plant residues (see review by Jenkinson and Ladd, 1981). These studies have shown that the decomposition process can be divided into two phases: an initial rapid phase, which results in loss of about two-thirds of the residue C in 1-2 years in a temperate climate, and a second phase in which decomposition follows a simple first-order kinetic model. The second phase has been followed for 10 years in Germany and for 7 years in Costa Rica with no evidence that first order kinetics will fail to describe the complete decay process (Sauerbeck and Gonzalez, 1977; Gonzalez and Sauerbeck, 1982). The decomposition pattern is not substantially affected by soil properties or crop residue type, at least within the ranges tested (Jenkinson, 1971), and patterns obtained in different climates can essentially be superimposed by using an appropriate rate conversion factor. Compared to results in England and Germany, the decomposition process was four and two times faster at sites in Nigeria and Australia (Jenkinson and Ayanaba, 1977; Ladd et al., 1981).

Knowledge of residue decomposition patterns can be used to predict the size of a pool of labile organic matter (LAB) in soils. The size of the LAB pool depends only on the particular decomposition pattern and the amount of residue inputs (Figure 1). It can be seen from Figure 1 that fairly small differences in the pattern of loss of residue C result in large differences in the size of the LAB pool. The LAB pool will contain partially decomposed plant and microbial residues, live microorganisms, and products of transformation (cf. Paul and Juma, 1981) and presumably contains much of Stanford and Smith's (1972) potentially mineralizable N. The LAB pool probably averages about a quarter to a third of the total SOM in temperate-region soils but would probably be less than this in tropical soils.

Figure 1. Measured decomposition of plant residues and calculated accumulation of a labile (LAB) pool of organic carbon under fallow field conditions for a range of soils and climates in Costa Rica. Figures in parentheses are final equilibrium values. Letters NA, PT, PA, etc., are different soils. (From Gonzalez and Sauerbeck, 1982.)

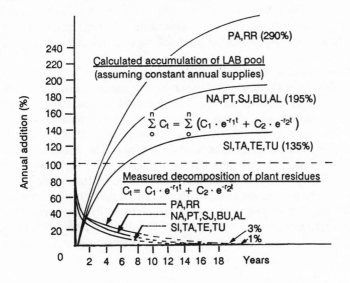

In most models the physical and chemical significance of the older organic fractions and their function as a source and a sink of nutrients is rather vague. For example, some authors seem to suggest that very old organic matter is recalcitrant because of its chemical nature alone, sometimes referring to this as chemical protection. In other cases the very old material is implied to be protected by chemical interactions with mineral colloids; some people call this chemical protection also, whereas others call it physical protection. In the models considered in this chapter, physical protection is more explicitly considered for the fraction of intermediate age rather than for the very old fraction. The conceptual or experimental basis for this is not apparent.

The chemical structure of organic molecules, by itself, is insufficient to account for the extreme variation in age and turnover times. Although humic molecules are undoubtedly more recalcitrant than natural biopolymers, their intrinsic chemical recalcitrance is much smaller than the observed stabilization of organic matter in soils. Old, humic fractions of soil organic matter, with ages in the thousands of years, have a half life on the order of weeks when extracted materials are added to unextracted soil. It is logical that physical-chemical interactions and protection, like chemical structures, would also be continuously, rather than discretely, variable. We propose that a logical basis for defining pools is the basis of scale. Colloidal or molecular level interactions among organic molecules and minerals, as well as the chemical nature of the organics, are hypothesized to stabilize the most protected, oldest organic components; this fraction could be identified as chemically or colloidally protected organic matter (COM). Structural or intra-aggregate level interactions, for example, organics sequestered in small pores, would lead to intermediate stability; this fraction could be identified as physically or structurally protected organic matter (POM). At least to some extent, the size of these pools would be controlled by different factors: the colloidally protected pool will be regulated by mineralogy and texture, which are invariable over time scales of interest, whereas the structurally protected pool will be controlled by tillage and soil disturbance (Table 1).

This scenario is consistent with the concepts of Tisdall and Oades (1982) and Oades (1984) on the role of different organic constituents in soil aggregation. Roots and fungal hyphae are considered to promote stabilization of macroaggregates, and as such macroaggregation will be controlled to a large extent by soil management practices, such as crop rotation and tillage, which influence the growth and decomposition of plant roots. The stability of microaggregates (less than $250 \mu m$) depends on persistent binding agents, such as complexes of clays, polyvalent metals, and organic matter (which includes polyaccharides and other organic polymers). The stability of microaggregates and their organic binding agents is more characteristically a property of the soil and is less dependent on soil management. Elliott (1986) used this model to demonstrate that N and P in microbial and organic materials, which are

Table 1. Pools of SOM and nutrients, generalized turnover rates, and hypothesized primary controls of pool size

Pool	Turnover time	Pool size controls
Unprotected		
BIO (microbial biomass)	2.5 yr ±, temperate 0.25 yr ±, humid tropics	Substrate availability
LAB (labile)	20 yr ± temperate 5 yr ±, humid tropics	Residue inputs, climate
Protected		
COM (colloidal protection)	1000 yr ±	Soil mineralogy, texture
POM (structural protection)	Depends on physical disturbance	Tillage and aggregate disruption, soil particle-size distribution

normally protected within macroaggregates in undisturbed soils, were mineralized by cultivation of North American grassland soils. Similarly, Skjemstad et al. (1986), using spectroscopic analysis of whole soil and density fractions, concluded that physical protection of SOM, rather than inherent recalcitrance, was the conservation mechanism in undisturbed Vertisols of subtropical Australia.

The nutrient content of the older organic fractions and their contribution to nutrient release is largely undefined except for N, and little information exists on this. Chemical fractionation of SOM gives fractions that vary in age and N content (e.g., Campbell et al., 1967). As the SOM components age they contain less N. If it is assumed that the various SOM fractions are at equilibrium, the rate of mineralization will equal the reciprocal of the age and the relative contributions of the fractions to N mineralization can be estimated on an annual basis. Such a procedure would be quite valuable if a biologically meaningful fractionation scheme could be devised, but it would not be generally applicable to agricultural systems because their SOM components are not at equilibrium.

A new approach to studying the turnover of older SOM components is to use differences in natural ^{13}C abundance levels in SOM arising from differential discrimination between ^{12}C and ^{13}C-CO_2 during photosynthesis in C$_3$ and C$_4$ plants. This method has been applied to SOM turnover in long-term experiments in which native C_3 plants were replaced by C_4 crop plants, e.g., forest by sugarcane in Brazil (Cerri et al., 1985) or vice versa, e.g., prairie by either wheat or timothy at Sanborn field at the University of Missouri (Balesdent et al., 1988). This approach may provide a means to experimentally test concepts of cycling of old C pools, but it does not directly give information on N, P, or S turnover.

Nutrient Pools in the Organic Matter of Tropical Soils

With regard to nutrient cycling, the effects of climate on SOM quality or lability and on turnover rates of organic nutrients are certainly of greater significance than are the effects of climate on total storage of SOM and nutrients. Few studies directly measure climatic effects on nutrient distribution among SOM pools, yet it is clear that the turnover times of unprotected pools, LAB and BIO, will be accelerated in warm, humid climates but not in dry tropical regions. This is a consequence of more favorable conditions for decomposition and greater C inputs in the humid environment.

Greater rates of soil N mineralization also indicate more rapid nutrient turnover. It has been concluded that net N mineralization, and subsequent nitrification, are greater in tropical forests than in temperate forests (Robertson, 1984; Jordan, 1985; Vitousek and Sanford, 1986; Vitousek and Matson, 1988); however, few tropical sites have been studied and there are large differences among sites, which seem to be related to differences in quantity and quality of litterfall (Vitousek and Matson, 1988). It is not clear what fraction of this increase is derived from mineralization of the large input of plant litter N, as opposed to mineralization of N in SOM.

We propose that the LAB pool is smaller in humid tropical soils than in comparable temperate-region soils. Continuously warm and moist soil conditions would result in more rapid and complete decomposition of unprotected organic materials. However, large and continuous litter input may compensate in part. The tendency for somewhat lower C/N ratios in the tropics, reported by Post et al. (1985), is consistent with more complete degradation of unprotected SOM. One predicted consequence of a smaller LAB pool would be a smaller fraction of the total organic N, P, or S mineralized under any given conditions (for example, in the Stanford and Smith [1972] assay).

The size and turnover of the oldest, colloidally or chemically protected pool (COM) should be relatively insensitive to moisture and temperature but highly sensitive to soil mineralogy and texture. The size of the COM pool will probably be larger in highly weathered, acid tropical soils where mineralogy is dominated by amorphous oxides of Fe and Al—that is, where areas of reactive mineral surfaces are large—than it is in temperate-region soils where mineralogy is dominated by layer silicate clays. However, consistent differences between temperate- and tropical-region soils would not be expected, other than through indirect effects of climate on mineralogy.

Dynamics of SOM in Tropical Ecosystems

Accelerated turnover of unprotected SOM in the humid tropics is of great importance for soil management and soil disturbance. Faster release of nutrients from existing LAB pools could lead to greater nutrient losses if sinks, particularly vegetation, are removed. Also, physical disturbance, and most dramatically intensive tillage, releases a portion of nutrients in the protected POM fraction to the LAB pool. The excellent aggregate structure of soils of the humid tropics suggests that in the undisturbed state POM pools may be at least as large as in temperate soils. Yet once protection is removed, losses of these nutrients should be faster in the humid tropics. In fact, tillage generally results in rapid release of organic N, P, S, and C in the humid tropics (Greenland and Nye, 1959). This explains, at least in part, the common observation that temperate-region Mollisols remain fertile much longer than do tropical-region Oxisols after initial cultivation. This is one sense in which it may be appropriate to speak of humid tropical soils as being more fragile than temperate-region soils.

Microbial Biomass as a Nutrient Source/Sink

The dual role of soil microbes as a catalyst and as a source/sink in nutrient transformations is now widely accepted. Direct measurement of microbial biomass indicates that 1 to 5% of the total organic C and N are stored in living tissue. The importance of the microbial biomass nutrient pool is further magnified by its more rapid turnover than occurs in total soil organic matter. Thus it constitutes a large part of the active fraction. At present, methods for direct measurement of microbial N, P, and S are somewhat less certain than are those for microbial C. The most widely used procedures for all these nutrients are based on the chloroform fumigation-incubation procedure developed for C by Jenkinson and Powlson (1976).

If the rate and extent of substrate degradation are assigned to the catalytic function of microbes, then the source/sink activity can be partitioned into the following factors:

a. The fraction of degraded C that is assimilated into new biomass.

b. The C/nutrient ratios of biomass.

c. Any mass changes associated with biomass maintenance.

d. The rate of cell death or turnover.

We have abundant data from pure culture studies and some soil measurements for the first two factors (a and b), although the extent to which these factors vary in soil and the controls on this variability are poorly defined. The provision for variation in biomass element ratios has been included in some recent models (van Veen et al., 1984), pointing to the sensitivity of mineralization and immobilization to this parameter. Differences in C-assimilation efficiency resulting from tillage systems have also been proposed to be important determinants of soil C storage (Holland and Coleman, 1987). The last two factors (c and d) are difficult if not impossible to investigate directly in soil. Knowledge of the physiology of microbial persistence and viability in natural environments is also very limited.

The current emphasis on microbes as a source and a sink for nutrients should not be taken to indicate that no questions remain to be answered about the catalytic role of microbes. Research in soil enzymology has concentrated on standard determinations of kinetic parameters and temperature response and usually has failed to address the significant questions of how synthesis of these enzymes is regulated, how their persistence is controlled, and how their actual function relates to potential activity (enzyme assays) and variable soil conditions. The energy cost for extracellular enzyme synthesis has not been defined. Microbial uptake systems are very well characterized in pure culture, but their significance is infrequently considered in natural systems. An example where this has been considered and is seen to be highly significant is with regard to uptake of different N sources. It is reasonably well demonstrated that NH_4+ is almost universally preferred over NO_3-, largely removing the latter from N turnover processes. But what is the situation with regard to NH_4+ and organic N sources? Two further questions related to the catalytic role of microbes are considered elsewhere in this chapter: the effects of nutrient interactions on degradation and the mechanisms by which organic substrates are protected from enzymatic attack.

There is no evidence, or reason to suggest, that the size of biomass nutrient pools will be any different in tropical than in temperate-region soils. Qualitative differences in the composition of both the microbial community and soil fauna—that is, differences in species composition—are likely, yet it is doubtful that characterizing such differences is the most fruitful approach to understanding differences in nutrient cycling. It is probable that the turnover rate of biomass nutrient pools will be greatly accelerated in moist tropical climates, particularly where primary productivity is high. There should be much more gross mineralization and immobilization in tropical soils.

These hypotheses are supported by calculations from Paul and Voroney (1983) based on data from cultivated soils in Canada, Brazil, and England (Table 2). Microbial C and N were lowest in Brazil, although not greatly different from the soil in England. Yet C inputs were approximately an order of magnitude higher in Brazil. This and the accelerated rate of decomposition caused estimated turnover times for soil C and microbial biomass to be approximately 10 times greater in the temperate than in the tropical-region soils. N flux through the biomass was thus also about tenfold higher in the Brazilian soil. Although this analysis is obviously limited with regard to number and type of sites, the differences in the controlling parameters, C input and C turnover, will apply to most comparisons of temperate and humid tropical conditions.

The consequences of accelerated biomass turnover and more gross mineralization and immobilization of nutrients would seem to depend on the

Dynamics of SOM in Tropical Ecosystems

Table 2. Amount and turnover rate of C and N for cultivated soils from three locations. (Adapted from Paul and Voroney, 1983.)

Determination	Temperate		Tropical
	England	Canada	Brazil
C inputs (mg ha^{-1} yr^{-1})	1.20	1.60	13.00
Estimated turnover of soil C (yr)[a]	22.00	40.00	2.00
Microbial C (kg ha^{-1})	570.00	1600.00	460.00
Microbial N (kg ha^{-1})	95.00	300.00	84.00
Estimated microbial turnover time (yr)[a]	2.50	6.80	0.24
Estimated N flux through microbial biomass (kg ha^{-1} yr^{-1})	34.00	53.00	350.00

a. Derived from simulation modeling

state of the system. When nutrient source-sink relationships are disturbed—for example, when vegetation is removed—there may be a greater potential for rapid nutrient release and loss. This would not be the case in undisturbed ecosystems or where continuous plant nutrient accumulation is maintained.

It is not clear how accelerated biomass turnover would affect the competition between plants and microbes for nutrients. It could be argued that accelerated gross immobilization would result in greater pressure on plants in terms of nutrient acquisition. Such selection pressure might favor plant-microbial symbioses as mechanisms of nutrient acquisition (mycorrhiza, symbiotic N-fixers). However, net microbial mineralization of nutrients is probably more significant in this regard than are microbial turnover rates. An alternative hypothesis is that the higher and less variable temperatures would make nutrient flux through the microbial biomass fast and constant. This would minimize the significance of microbial biomass as a competitive nutrient sink in tropical soils. Perhaps an indication of this would be provided by the relative nutrient pool sizes of microbial and plant biomass in different climates.

Differences in Cycling among N, P, and S

While the bulk of the N and S in most topsoils, regardless of climate, is held in organic molecules, inorganic forms of P may constitute a large fraction of the total. As soils weather, the significance of organic P as a nutrient source increases. As a generalization, it appears that the significance of P as a soil fertility problem is greater in tropical soils than in temperate-region soils (Vitousek, 1984). Despite the obvious importance of P mineralization and immobilization, much less is known about its behavior than about N.

Sanchez (1976) suggests that a wide C/P ratio is a symptom of P deficiency, but this concept is not universally accepted. Soil organic C/P ratios in tropical soils are highly variable but are not necessarily wider than those of temperate region soils (see preceding discussion). The greater variability of C/P than C/N ratios is often interpreted to mean that P mineralization is uncoupled from C and N mineralization. The much greater susceptibility of soil organic P to alkaline extraction, relative to other elements, also indicates that its storage is somewhat independent of C and N. Since P occurs almost entirely in the ester form (C-O-P) while N is covalently bonded directly to C, it is plausible that their behavior differs. Because S occurs as both C-S and C-O-S, its behavior may involve characteristics of both. McGill and Cole (1981) and Hunt et al. (1983) have used this concept of different bond types to account for interactions among elements and differences in their behavior. It was proposed that the mineralization of esters is regulated by demand for the nutrient, but mineralization of elements covalently bonded to C is regulated by the factors that control use of the energy-yielding substrate.

It is clear that soil organic P can be hydrolyzed extracellularly and mineralized without C degradation, but it may also be assimilated and mineralized with concomitant oxidation of the C. And while the common idea of N mineralization includes microbial uptake followed by oxidation of the C, it is also true that extracellular deaminases exist in soil. In neither case, P or N, has it been conclusively demonstrated what the balance is between extra- and intracellular hydrolysis. More important, the regulation of these processes in soil is poorly documented. For example, how much effect does P, N, or S availability have on mineralization of the same or another element? How variable are the element ratios of microbial biomass and how are these regulated? What is clear is that mineralization-immobilization processes for P and S may differ significantly from those for N and that interactions among these elements need to be considered.

Research Needs

At the heart of any discussion involving differences in climatic conditions is the question of temperature and moisture regulation of soil processes. Describing and understanding these and other stress effects on microbes is in some respects the central problem. Although single-factor response studies (effect of temperature on ..., effect of soil moisture on ...) have been done ad infinitum, we are still not certain how to integrate these to quantitatively predict their effects on mineralization and immobilization of nutrients (or most other processes). This problem will not be resolved simply with multiple-factor experiments (effects of temperature and moisture on ...), but will require greatly improved knowledge of how microbes in soil respond physiologically to various forms of stress.

Dynamics of SOM in Tropical Ecosystems

We propose that all the organic matter in soils is readily mineralizable and that associations within soils, at both the molecular and aggregate levels, prevent this from occurring. Much more attention needs to be given to (a) the chemical nature of associations of organic matter in soils, both between organic molecules themselves and between organic and inorganic soil constituents, (b) the effects of different types of associations on the stability of SOM components to biological decomposition, (c) how soil and cropping management systems influence interactions between organic and inorganic soil constituents, and (d) how the overall architecture of the soil system affects decomposition processes. These studies could be especially fruitful in the tropical soils that have reasonable to high clay contents and exceptionally good physical structure, that is, Oxisols.

Characterizing organic nutrient quality or lability remains a problem, more so in soil organic matter than in plant residues. The chloroform fumigation technique is approaching the status of a universally accepted standard for biomass C measurement. Also, this approach is being widely used for biomass N, and to a lesser extent for P and S. Yet questions remain about its application: What is the most appropriate control? What is the best conversion factor? How do fresh residues and substrates affect the results?

Although the assumptions used in measuring active pools (e.g., Paul and Juma, 1981) can be questioned, such approaches do provide information about the activity of organic nutrients. These tests must be functional, that is, based on biological incubations. It is important to establish widely applicable approaches, if not standard techniques. Although non-tracer techniques, such as the Stanford and Smith (1972) determination of potentially mineralizable N, are of significant value, this area rather than fertilizer efficiency studies is the appropriate place for application of N-isotope studies in tropical soils.

Much less is understood about mineralization and immobilization of P and S than about N. Interactions among elements, such as the effect of P limitation on N storage and cycling, are also poorly described. The significance of P limitation in many tropical soils indicates that these are important problems for future research.

The obvious complexity of mineralization-immobilization makes this an excellent candidate for useful application of mathematical modeling. Yet, a division generally persists between those who model and those who experiment on mineralization and immobilization. There is a need for joint effort between experimenters and modelers to produce models that will both stimulate the development and testing of new hypotheses and which have value at a practical agricultural level.

Management of Soil Organic N, P, and S
Nutrient Cycling in Natural and Agricultural Ecosystems

The soil organic matter content of a given soil in a natural ecosystem approaches a steady state when inputs from plant production balance the decomposition activities of soil microorganisms and soil fauna. The steady state content of C and N attained is governed by the balance between inputs and decomposition rates and the capacity of the soil to protect SOM from decomposition. Although native SOM contents vary widely, they rarely exceed 5-6% except under conditions where microbial decomposition is limited by reduced aeration or where organic matter protection by interactions with minerals is exceptional. For example, with allophane in volcanic soils (Sanchez, 1976; Paul, 1984; Stevenson, 1986). The organic matter content of undisturbed grassland and forest soils changes very slowly; where large changes occur naturally they are usually associated with fire or major climatic changes that affect vegetation. In natural ecosystems, internal recycling of nutrients between primary producers, consumer and decomposer organisms, and abiotic storage pools greatly exceeds flow through the system from atmospheric inputs or leaching losses (Crossley et al., 1984). Agroecosystems differ from natural ecosystems in that energy fluxes, nutrient cycles, and hydrologic characteristics are regulated to varying degrees by physical manipulation of the soil and external inputs of water, nutrients, and energy. The overall impact of these management technologies is disturbance of the normally conservative nutrient cycles and accelerated release of nutrients in SOM to the soil abiotic environment where they can either be taken up by plants or lost through leaching or volatilization. Soil organic matter contents decline until a new steady-state, characteristic of the particular system, is reached. The new steady-state SOM content would be reached fairly quickly in tropical soils if the system were controlled only by the intensity of decomposition processes; however, the rate of release of protected organic matter will also be a major factor because most of the change in SOM levels will be due to changes in the size of this pool.

An important difference in nutrient cycling between natural and agricultural ecosystems is the relative synchrony between plant growth and microbial activity. In natural ecosystems the growth of microorganisms and plants is often in synchrony because they have the same general environmental requirements for growth. Natural selection and species diversity enables biological activity to occur over much of the year, subject only to moisture limitation in the tropical environment. Most available nutrients are used by plants or microorganisms and are recycled through the various SOM pools. Soil contents of available nutrients are minimal during periods of high biological activity (although nutrient fluxes may be high) and nutrients are therefore not susceptible to loss by leaching or volatilization during these times. In agricultural systems, high crop demand for nutrients occurs during defined time intervals, and there may be considerable periods of time when plants are absent. If release of nutrients is insufficient to meet crop demand, supplementary fertilizer may be needed. In the same system, nutrients mineralized during periods when crops are not growing will accumulate in soil and, together with residual fertilizer, are subject to loss by leaching or volatilization. Even with continuous cropping the root systems of annual crop plants require some time to develop and are never as extensive as those of perennial plants; hence, annual crop plants are unlikely to be as effective at recovering nutrients from soils as perennial crop plants or natural perennial vegetation. Consequently, many agricultural ecosystems are inherently more nutrient-leaky than are natural ecosystems. More attention should be given to including the natural recycling capacity of perennial plants in agricultural systems. Agroforestry systems are an example of where this issue has been addressed.

One approach (McGill and Myers, 1987) to predict synchrony between N mineralization and crop uptake of N is shown in Figure 2. Here N mineralization rate is assumed to depend only on soil moisture and temperature regimes, which are combined into a single index. On this basis synchrony is predicted to be reasonable for crop growth in temperate North America but N mineralization both precedes and continues beyond crop growth. In subtropical and tropical environments, where more than one crop may be grown, predicted synchrony varies considerably, suggesting that there may be a need to add fertilizer N at some times and to conserve mineralized N at other times. This analysis does not include factors that alter the availability of protected SOM pools, but it identifies a probable need to develop nutrient conservation strategies. Evaluations of this kind illustrate the potential utility of even simple models and also the need to determine actual N, P, and S mineralization rates under field conditions.

Management of Nutrient Storage and Release from Different SOM Pools

The Labile Organic Pool

As a first approximation, the size of the LAB pool will be controlled by the quantity of residue inputs and climate. Several reasonably long-term agronomic field experiments demonstrate that, as predicted, the size of the LAB pool is directly proportional to residue inputs. C retention in an 11-year study in Iowa (Larson et al., 1972) was 21% of that added at various residue addition rates up to 16 Mg ha^{-1} yr^{-1} with no effect of crop residue type (among alfalfa, corn stalks, oat straw, and bromegrass) on C retention. In a 20-year experiment at Sameru, Nigeria, M. J. Jones (1971) found C retention of 14% and N retention of 30% for additions of farmyard manure up to 12.5 Mg ha^{-1} yr^{-1}.

More research may well show that other variables will affect residue decomposition patterns and hence the size of the LAB pool, but the general principle that the size of this pool is proportional to the amount of residue added

Figure 2. Effect of climate and cropping on synchrony between relative crop yield and potential for mineralization of soil N based on a soil moisture x temperature index. (Adapted from McGill and Myers, 1987.)

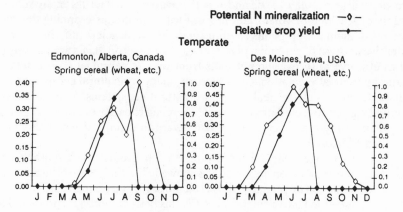

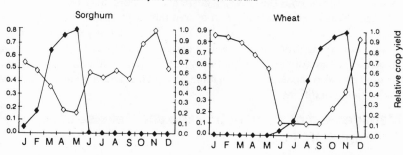

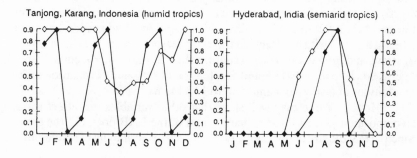

will hold. Factors likely to affect the decomposition pattern and C retention include the physical state of the residue, where it is placed in soil, and the method and frequency of tillage. Residue chemical composition and soil variables that affect the activity of decomposer organisms may also be important. Animal manures are 3 to 4 times more effective than plant residues at increasing the LAB pool because they have already undergone the rapid initial decomposition phase.

In absolute terms, the size of the LAB pool derived from plant residue additions will be smaller in tropical soils than in temperate-region soils, where a value of about 2.5 times the annual residue addition rate has been established (Sauerbeck and Gonzalez, 1977). Using this as a maximum value, and with constant annual residue inputs of 1000 kg C ha^{-1}, the LAB pool would contain a maximum of 2500 kg C ha^{-1}, or 250 kg N ha^{-1} and 25 kg P ha^{-1} if the C/N and C/P ratios are 10:1 and 100:1, respectively. It also responds quickly to changes in residue inputs so that current agricultural management is important.

One question on which there is little information is whether the nutrient content of the LAB pool is influenced by the amounts of nutrients in added residues or fertilizers. When various crop residues containing approximately equal amounts of C but different amounts of N, P, and S were incorporated annually into a soil cropped to maize, soil organic C, but not N or S levels, were increased by essentially identical amounts after 11 years (Larson et al., 1972). Soil organic N and S levels appeared to increase with increasing content in the residue, although incomplete residue analysis and fertilizer addition complicate this interpretation. Residue addition had a relatively small effect on soil organic P content; for example, the addition of 16 Mg of alfalfa ha^{-1} yr^{-1} caused soil organic C to increase by 48%, organic N by 41%, and organic S by 49%, whereas soil organic P only increased by 14%. Besides reinforcing the concept that soil organic P behaves somewhat independently from C, N, and S, this experiment strongly suggests that the nutrient content of crop residues will affect the nutrient content of the LAB pool. Fertilizer additions may have a similar effect, but we are not aware of data that would demonstrate this for N or S; for P, however, long-term additions of fertilizer P (35 kg P ha^{-1} yr^{-1} for 125 years) increased the organic P content of SOM in the Park Grass experiment at Rothamsted, England, by 1.5 times without altering SOM contents (Hawkes et al., 1984).

Protected Organic Matter Pools

The size of the LAB pool of SOM and the controls over this pool are such that major changes in SOM contents cannot be achieved by manipulation of this pool alone. In fact, the large declines in SOM associated with tillage of soils and the gains in organic matter when tillage is stopped are mostly associated with the POM pool although turnover of the LAB pool may also be affected. The most effective way to increase SOM in cultivated soils is to stop tillage. This, of course, immediately reduces the availability of nutrients to crops but results in gradual accumulation of nutrient reserves in SOM. Shifting cultivation in the tropics operates on alternating phases of exploitation and accumulation. In effect, similar systems are widely practiced in the temperate regions on a compressed time scale through the use of short-term crop rotations. The most effective rotations are those in which a forage legume is grown to accumulate N via biological N fixation in the no-till phase of the rotation. This system leads to rapid accumulation of N and hence to a short-term rotation.

Although tillage is in many ways the most powerful tool available for the manipulation of storage and release of nutrients in SOM, its effects on the soil system are complex. Tillage and crop residue management practices have a large influence on soil temperature, water, and aeration regimes and on the spatial and temporal availability of energy and nutrients to microorganisms. The redistribution of organic matter and soil organisms with reduced tillage is a major factor responsible for greater recycling and retention of N than with conventional tillage (House et al., 1984; Fox and Bandel, 1986). Organic matter contents of the upper layers of soil, microbial biomass, and reserves of potentially mineralizable N are often significantly higher with no-tillage than with moldboard plow tillage (Table 3). Such increases in microbial biomass and activity and organic N reserves result from surface placement of residues, accumulation of SOM, and a more optimal water status for biological activity in the surface 0-10 cm of reduced-tillage soils in both temperate and tropical regions (Ayanaba et al., 1976; Doran, 1987a). Increased microbial biomass is also associated with increases in plant root activity near the surface of no-tillage soils (Lynch and Panting, 1980; Carter and Rennie, 1987).

Table 3. Effects of tillage on soil water content, chemical components, and soil microbial biomass as a function of soil depth at six (four continuous maize, two wheat/fallow) long-term (6-13 year) experiments in the United States. (Data adapted from Doran, 1987a)

Soil parameter	Ratio of values for no tillage to moldboard plow tillage			
	0-7.5 cm	7.5-15 cm	15-30 cm	0-30 cm
Water content	1.28*	1.08	1.08	1.13
Water-soluble C	1.47*	0.98	1.24	1.23
Total organic carbon	1.42*	1.00	0.94*	1.06
Total Kjeldahl N	1.29*	1.01	0.97	1.06
Potentially mineralizable N	1.37*	0.98	0.93*	1.05
Microbial biomass	1.54*	0.98	1.00	1.13

* Significant difference (p < 0.05) using F-test for tillage across locations

The magnitude of management-related changes in the properties of tilled layers and microbial responses can greatly depend on previous management, cropping, and degree of tillage. The levels of organic C, N, and microbial biomass in surface soil of a winter wheat (*Triticum aestivum* L.)/fallow rotation in western Nebraska were inversely related to degree of soil tillage during fallow (Doran, 1987b). In previously cultivated land where initial SOM levels were lower, these differences were much less pronounced than where tillage comparisons were initiated in native grass sod. Also, over an 11-year period, the total soil N content with no-tillage management was 9% greater than when crop/fallow was first initiated. In converting from grassland to wheat/fallow, declines in soil organic C and N levels, regardless of tillage management, reflect decreased inputs of C and N resulting from reduced plant production and residue inputs. In either case, however, reduced tillage has conserved more N in the upper layers of soil — through reducing net mineralization of crop residues and SOM — than has subtillage or plowing.

Tillage-induced differences of N in the upper layers of soil reserves of potentially mineralizable N and microbial biomass may vary with climate and cropping management practices. Differences in mineralizable N reserves between plow and no-tillage management at six long-term experiments across the United States ranged from 12 to 122 kg N ha^{-1} and were highly correlated with mean annual precipitation (Doran, 1987b). These trends likely result from increased cropping intensity and plant productivity associated with increasing rainfall. Differences between tillage management were least and values for potentially mineralizable N lowest for wheat/fallow production in a low-rainfall region. Higher contents of mineralizable N and greater differences were observed at four locations under continuous cropping with maize, especially at the most humid location where a rye cover crop was also planted.

Interactions between microbial activity and mineralization of soil organic N are often controlled by environmental factors. Predicting how changed environmental conditions in reduced tillage soils will affect net mineralization is difficult because the contrasting effects of increased water and reduced temperatures on net mineralization may vary during the growing season and across climates (Fox and Bandel, 1986; Doran and Smith, 1987). Also, increased microbial biomass levels in no-tillage surface soils during the growing season can serve as a sink for immobilization of N. Greater immobilization of fertilizer N has been observed in no-tillage crop production than in production using plowing or shallow tillage (Rice et al., 1986; Carter and Rennie, 1987).

The effectiveness of tillage in releasing the N contained in soil microbial biomass and organic N reserves is also influenced by soil type and the density of plant roots. The productivity of grass pastures in subtropical Australia is often limited by reduced availability of mineral N as a result of accumulation of root and plant debris with a high C/N ratio and increased immobilization of N in microbial biomass. Periodic cultivation of grass pastures increases mineralization of soil N and stimulates grass production through changes in root density and mineralization of microbial and organic N reserves (Table 4). In clay soils the N mineralized by cultivation may come largely from stabilized forms of organic N, whereas in sandy, loamy, coarse-textured soils microbial biomass may be the predominant source of mineralized N. Changes in microbial biomass resulting from cultivation paralleling those for root biomass suggest an association between changes in root density and microbial biomass levels in soil.

Table 4. Effect of soil cultivation on N budgets for the 0-30 cm soil depth interval of grass pastures at two sites in Queensland, Australia. (After Doran, 1987b.)

Plant or Soil Component	Green panic, clay (kg N ha^{-1})			Buffelgrass, sandy clay loam (kg N ha^{-1})		
	No tillage	Chisel plow	difference	No tillage	Plow/ resown	difference
Plant nitrogen						
Tops	45	62	+17	28	41	+13
Roots	84	102	+18	116	94	-22
Soil nitrogen						
Nitrate + ammonium	4	10	+6	8	7	-1
Mineralizable N	945	882	-63	480	459	-21
Microbial biomass N	318	332	+14	153	116	-37
Total organic N		8143			3144	

Research Needs

Agricultural management systems control residue placement, soil disturbance, and the *in situ* production of soil organic matter. The resulting effects on soil physical properties and substrate availability greatly influence plant and microbial activities and the cycling of C, N, and other nutrients. Interactions between the soil physical, chemical, and biological characteristics with various systems of organic matter management are greatly influenced by climate, soil, and initial SOM levels. Development of appropriate management strategies for most efficient utilization of N and other elements requires better understanding of these interactions, especially in tropical ecosystems where little research of this nature has been previously conducted. Suggested research needs are to:

1. Determine the short- and long-term effects of tillage and crop management systems on organic nutrient reserves in tropical soils. In particular, what role do surface placement and lack of tillage play in the accumulation of occluded organic matter (POM) and nutrient reserves? Will immobilization of nutrients in POM decrease with time in undisturbed soils, resulting in more recycling and less nutrient tie up?

2. Identify the processes responsible for the tillage-induced declines in SOM. The continual decline over many years, coupled with several observations that sieved virgin soils and soil cores mineralize N at the same rate, suggests that soil disruption by the tillage instrument is not the primary driving force. Knowledge of the mechanisms involved in the tillage effect might lead to more controlled nutrient-release practices, may aid the design of tillage tools, and may provide a basis for choosing between alternative tillage methods. A possible cause for the slowness of the tillage effect is the gradual exposure of new soil to the more highly physically disruptive environment at the soil surface. The central issue may prove to be one of scale, i.e., tillage equipment operates at a macroscale whereas the forces generated, for example, by drying and rewetting act at a microscale level.

3. Better define the function of soil organic matter. Both biological and physical-chemical functions, need to be considered. What is the optimum balance between exploitation of SOM as a source of nutrients and energy and its conservation as a conditioner of the soil physical environment? More emphasis on comparisons of virgin and cultivated soils may provide insight into mechanisms of changes resulting from agricultural management.

4. Include several elements in most studies on SOM. Much of the work on organic matter dynamics, for example has been limited to C and application to N, P, and S is by inference. Almost all the work on the effects of tillage on nutrient release and storage has dealt only with N. Future studies should include several elements to establish similarities, differences, and interactions among and between them. Such information is likely to influence decisions on management of SOM.

5. Better define the concept of a biologically inert pool of organic matter, i.e., the COM pool. This concept is intuitively unsatisfactory. What could be the mechanisms that so completely protect inherently decomposable organic molecules from decomposition? Surely there can be no one mechanism because soil inorganic components vary so greatly as a function of weathering; or, are overriding interactions with Al and Fe species common to all soils?

Organic Matter and Cationic Nutrients

Mechanisms of Interaction

Charge development on SOM is pH dependent and is predominately negative. The principal functional groups involved in negative charge development are carboxylic and phenolic acids. Some positive charge development through protonation of amino groups can also occur, but it is quantitatively small compared to negative charge development. Interaction of cations with SOM involves several different bonding mechanisms. Basic nutrient cations (Ca, Mg, and K) interact predominantly via electrostatic attraction, whereas divalent transition metal nutrients (Zn, Mn, Fe, and Cu) and trivalent Fe form coordinate linkages of varying strength with SOM. Nonnutrient cations, especially Al, will compete for coordination sites. Organic matter-metal complexes may be monodentate, bidentate, and so on, and they may include the formation of cyclic chelate structures. Studies of cation binding to SOM are usually carried out on organic matter extracted from soil and fractionated into the traditional fulvic and humic substances. The strength of binding of various metals to soil humic substances usually follows the order Fe (III) > Al > Cu > Ni > Co > Pb > Zn > Mn > Ca > Mg, but the order somewhat depends on pH. The strength of binding of a given metal also depends on the amount added, indicating that a range of organic complexing sites, with different affinities for the metal, are involved.

Importance of Organic Matter/Cation Interactions in Tropical Soils

Cation Exchange and Base Retention

The CEC of soils is especially important for retention of the basic cations (Ca, Mg, and K) that interact with soil surfaces via electrostatic attraction. Both soil minerals and SOM contribute to CEC in soils; however, as soils are weathered, their CEC drops due to changes in mineralogy from 2:1 type layer aluminosilicate minerals to kaolinite and amorphous oxides of Fe and Al. The CEC of most soils dominated by 2:1 layer minerals is usually within the range of 15-30 $cmol_c$ kg^{-1} soil, whereas that of soils dominated by kaolinite and amorphous oxides is almost always less than 5 $cmol_c$ kg^{-1} soil and most of that may be associated with SOM rather than with the mineral components. Because there are large areas of highly weathered soils in the tropics, maintenance of SOM to provide CEC is more important in tropical than in temperate regions.

Approximate average values for the total acidity of extracted soil humic substances are between 700 and 1000 $cmol_c$ kg^{-1} organic matter (Schnitzer, 1978), indicating a high potential for SOM to add CEC to soils. However, much of this charge is not expressed in soils, both because it is pH dependent and because many of the negatively charged sites are blocked by interactions with

Al and perhaps also Fe. The amount of CEC contributed to soil by SOM is best measured by the multiple-regression technique used by Helling et al. (1964) since this avoids extraction of organic matter from soil. In soils from Wisconsin, SOM contributed about 180 cmol$_c$ kg^{-1} organic matter at pH 5 and changing pH altered CEC by about 30 cmol$_c$ kg^{-1} organic matter per pH unit. The maximum charge development of about 350 cmol$_c$ kg^{-1} organic matter (at pH 8.2) was considerably below that expected from *in vitro* measurements with soil humic substances and indicates the importance of nonexchangeable interactions with soil Al and Fe. Similar methodology has not been applied to highly weathered soils, but it is likely that the CEC of organic matter in such soils will be lower because a greater degree of blockage of negatively charged sites by Al and Fe can be anticipated. Relationships between CEC and SOM content for highly weathered soils of the Brazilian Cerrados were found to depend on soil pH (Lopes and Cox, 1977). SOM was only effective at increasing CEC levels above a pH of 5.5, which is consistent with blockage of exchange sites by either Al or Fe at lower pH values. This result suggests that attempts to increase the CEC of acid soils by increasing SOM levels may meet with short-term success but long-term disappointment as the soil system equilibrates to block newly formed exchange sites. Perhaps the approach that should be taken would be to raise soil pH by liming, but it is not clear that this will unblock previously blocked sites.

It is often considered that one benefit of the older, more biologically inert pools of SOM is that they provide CEC to soils; however, it is possible that these pools are also fairly inert on a chemical basis and contribute little to soil CEC. If the LAB organic pool were to be the most important organic source of CEC in tropical soils, maintenance of this pool at high levels would be an important objective for maintaining CEC.

Despite an incomplete understanding of the relationships between SOM and soil CEC and how to manipulate these relationships in acid soils, it is clear that destruction of SOM will reduce CEC levels and have a negative impact on base cation retention.

Micronutrient Availability

Organically complexed micronutrient metals, and perhaps also borate complexes with soil carbohydrates, are generally considered an important component of the labile reservoir of these elements in soils. In addition, soluble organo-metal complexes are often a major proportion of the micronutrients in soil solution and aid in the transport of micronutrients to plant roots. This is most important in high pH soils because, with the exception of B and Mo, the solubility of inorganic forms of all of the micronutrients decreases with increasing pH. Micronutrient deficiencies are common in tropical regions that have high pH soils which have been cultivated for long time periods, such as India, and this situation would likely be improved by increasing SOM contents.

Indirect Effects on Nutrient Availability and Soil Fertility

Complexing of organic matter with reactive Al and Fe surfaces reduces soil CEC but has beneficial effects in that it blocks these sites and reduces the capacity of soil to fix phosphate and sulfate. Increasing SOM may also stimulate desorption of phosphate and sulfate by acting as a competing anion. However, this may not be very useful if the inorganic anions are simply replaced by organic esters of these same elements. An unequivocal effect of complexing of Al by SOM is that it reduces exchangeable Al levels, which is important because Al toxicity is a common problem in acid Oxisols and Ultisols. Strong complexing with Al and Fe is probably also the reason that Oxisols and Ultisols have such good physical structure.

Overall, there are good and bad sides to the complexing of SOM with soil inorganic constituents but reduction in SOM contents has negative effects on all of the soil properties discussed in this section.

Research Needs

The case for developing a better understanding of SOM interactions with soil mineral components has already been stated in terms of their effects on nutrient release by mineralization processes. It is also important that such studies include an analysis of how interactions involving the various organic pools affect the contribution of these pools to cation retention and availability in soils.

Organic Sources of Nutrients and Crop Yield

The importance of SOM and organic residues to crop production has diminished in developed countries over the last 40 or so years with the availability of cheap energy. In these countries, chemical fertilizers are the main source of the major plant nutrients, and other practices that help to overcome adverse effects associated with lowered SOM contents have been adopted – for example – drainage and irrigation to correct poorer soil physical conditions. High-energy input agricultural systems are not, however, economically viable in many of the developing nations of the tropics where agriculture relies more on SOM, animal manures, and crop residues to supply needed nutrients.

Chemical fertilizers, in fact, are generally needed to supply sufficient N, P, K, and sometimes other nutrients to reach the yield potential of improved crop varieties. The slow release of nutrients from SOM is often considered advantageous (Sanchez, 1976; Avnimelech, 1986) for both soluble nutrients, such as N, and for nutrients that may interact with soil surfaces and become less available, such as P. It would seem that one advantage of chemical fertilizers over organic nutrient sources, however, is that nutrient forms, amounts, placement, and timing of addition are easily regulated and can be matched to the needs of a particular crop. Given best management practices, one would expect greater crop recovery of relatively mobile nutrients, such as N, with chemical fertilizers than with organic nutrient sources and this is generally so. It is less clear whether nutrients that chemically interact with soil surfaces, such as P, are best supplied by organic nutrient sources or by inorganic fertilizers.

A long-standing question is whether or not crop yield potential is affected by SOM level. Although most agronomists would consider that this is not the case for the main grain crops of the world, there are several recent studies that show higher yield potentials with organic manures than with inorganic fertilizers. For example, farmyard manure gave higher yields of both small grain and root crops than were achieved with fertilizers in the long-term experiments at Rothamsted and Woburn, England (Johnston and Mattingly, 1976; Cooke, 1977) and of maize in Colorado (S. R. Olsen, 1986). Similar results have sometimes been reported for paddy rice with various organic nutrient sources, including farmyard manures, green manures, and composts (Chatterjee et al., 1979; Tiwari et al., 1980; Guar, 1984; Kumazawa, 1984; Oh, 1984; Bhatti et al., 1985; Bouldin, 1988). Positive effects of organic nutrient sources seem to be most often observed with crop varieties that have high yield potential. Several of the authors cited also report better efficiency of nutrient use when organic nutrient sources are used in combination with inorganic fertilizers. The focus of most studies has been on N supply, and it has been suggested that the beneficial effects of the organic nutrient sources are due to their increasing the availability of P, K, or micronutrients (Avnimelech, 1986; Bouldin, 1988) or to better root development (Kumazawa, 1984). The release of N from organic sources late in the growing season is also considered important for high yields of rice, and S. R. Olsen (1986) believes that the supply of ammonium from decomposing organic matter provides a balance of ammonium to nitrate which results in better crop growth.

While there is ample evidence to suggest that organic amendments to soil can improve crop yield, this result is not routinely found; in fact, it is well known that crop yield can be depressed if there is net immobilization of nutrients during microbial decomposition of organic residues. Management of organic manures in flooded soils is not straightforward, and loss of nutrients from both the soil and the residue and/or generation of low molecular weight organic acids toxic to plants can occur. In some of the studies in which organic nutrient sources proved to be superior to inorganic fertilizers, it was suggested that the effect was due only to better supply of nutrients from the organic matter. Such a claim is extravagant as it is impossible to monitor all the effects of the organic matter on nutrient availability let alone on the soil-plant system as a whole. Besides effects on nutrient supply, which can be direct (nutrients supplied by the added organic matter) or indirect (effect of the added organic matter on the availability of soil nutrients), the organic matter will presumably also have varying effects upon soil physical properties and microbial and animal life in the soil. The experience with organic manures can to some degree be extrapolated to SOM, but with so many interacting factors in the soil-plant system it is clearly difficult to conclusively establish cause-and-effect relationships between either organic manures or SOM, nutrient supply, and crop yield. As noted out by Bouldin (1988), it is important to separate effects due to organic matter per se from those associated with its decomposition.

Research Needs

While decomposition of SOM and organic manures promotes plant growth through release of nutrients, microbial activity will have additional, less easily definable, impacts on the soil-plant system. Assessment of these additional impacts is essential not only to interpretation of experiments in terms of nutrient availability but also to attaining an overall understanding of the behavior of soil-plant systems. Without this understanding, management choices will be made on an empirical rather than a mechanistic basis and will have reduced success.

Conclusions and Recommendations

Soil organic matter impacts nutrient availability via active (biological transformation and mineralization) and passive (chemical) processes. Our knowledge of the dynamics of SOM and its associated nutrients has developed considerably in recent years. The concepts of pools of organic matter differing in susceptibility to biological decomposition, each responding differentially to soil, environmental, and agricultural management variables, has provided the foundation for understanding how SOM functions as a source and a sink of plant nutrients and how this function can be managed. Nevertheless, while concepts and models are in place, we have a limited capacity to directly measure the various organic matter pools and a limited understanding of the linkages between pools, In addition, we lack experimental verification for proposed mechanisms of protection of organic matter in soils.

Although the focus of future research should be on the dynamic aspects of nutrient availability, we must not forget the passive functions of SOM, which should also be evaluated in terms of the "pool" concept. For example, if the contribution of organic matter to soil CEC is largely associated with the old, stable organic matter pool, there is probably little opportunity for management of this property. If, however, the labile organic pool contributes significantly to CEC, maintenance of this pool, which is strongly affected by recent management practices, would be an important goal.

Differences in soil composition arise as a function of weathering of soils and these, coupled with climate, have a marked effect on the content and behavior of organic matter in soils. Consequently, in a general sense, there are differences in the behavior of SOM in tropical and temperate-region soils because of differences in weathering of soils and climate. At the same time, the biological, chemical, and physical principles governing the behavior of SOM are the same for all soils, and seemingly disparate soils may show remarkable similarities in a particular property; an example of this is the relationship between soil inositol phosphate content and phosphate adsorption capacity.

Because the scope of this chapter is rather wide, we have identified specific research needs at the end of each section. More general recommendations for future research follow.

Theme Imperative: Define the processes controlling nutrient release and storage in SOM and their regulation by soil, climatic, and management variables.

Research Imperatives:

1. Undertake comparative studies of nutrient cycling at sites carefully selected for variation in temperature and moisture regimes, soil mineralogy, native vegetation (natural ecosystems), and agricultural management.

2. Initiate long-term (decade scale), *in situ* isotopic tracer studies, using multiple labels to probe C and nutrient dynamics of labile pools and interactions between labile and more stable organic matter pools.

3. Because of the significance of phosphorus limitations to plant growth in the humid tropics, give high priority to determining the forms, interactions, and biological transformations of soil organic phosphorus.

4. Develop better ways of describing the quality or lability of nutrients in soil organic matter. Ensure that conceptual pools are meaningful in terms of biological activity and are experimentally measurable. Give more attention to extending laboratory-measured properties, such as N mineralization, to *in situ* nutrient behavior.

5. Define mechanisms of protection of soil organic matter and nutrients from mineralization and quantitatively relate them to soil chemical and physical properties and environmental variables.

6. Evaluate the biotic and abiotic regulation of microbial assimilation and release of nutrients and their relationships to soil nutrient dynamics.

Chapter 3

Interactions of Soil Organic Matter and Variable-Charge Clays

J. Malcolm Oades, Gavin P. Gillman, and Goro Uehara

with Nguyen V. Hue, Meine van Noordwijk, G. Philip Robertson, and Koji Wada

Abstract
Many tropical soils are dominated by variable-charge minerals, and the presence of organic matter, itself a variable-charge material, can strongly affect the overall charge properties of a soil and thus its ability to retain ions against leaching. Net charge can be negative or positive, depending on soil pH relative to the pH at which soil cation exchange capacity (CEC) and anion exchange capacity (AEC) are equal – that is, at the point of zero net charge (PZNC). The PZNC is a function of the various mineral and organic constituents of a soil and the degree to which each of the individual components imparts or impedes the expression of variable charge.

   Organic molecules interact in complex ways with soil hydrous-oxides and oxyhydroxides to affect soil charge properties. Organic cations, humic substances, and natural organic/hydrous-oxide complexes can affect (1) electrochemical properties (CEC, AEC), (2) soil pH and acidity, (3) specific adsorption of polyvalent ions such as phosphate and sulfate, and (4) soil physical properties such as aggregate stability and hydraulic conductivity. Clearly the degree to which soil organic matter affects or imparts variable charge is an important aspect of organic matter management in tropical soils.

We identify six major research areas deserving immediate attention. These are (1) identifying mechanisms for maintaining organic matter in variable-charge soils, (2) characterizing the organic matter pools that affect variable charge in soils, (3) quantifying the relationships between soil physical properties and surface charge in variable-charge soils, (4) defining the role of organic matter as a regulator of soil acidity and aluminum/manganese phytotoxicity in variable-charge soils, (5) quantifying changes in variable-charge properties brought about by soil disturbance, and (6) developing management strategies consistent with specific properties of variable-charge soils.

Working in Indonesia, Mohr (1930) categorized soils as either juvenile, mature, or senile. Mohr's senile soils were called lateritic soils and latosols by his contemporaries. They are now named Ferralsols and Acrisols in the United Nations Food and Agriculture Organization Soil Legend (FAO-UNESCO, 1974) and Oxisols and Ultisols in the United States Department of Agriculture Soil Taxonomy (Soil Survey Staff, 1975). As the nomenclature implies, Ferralsols or Oxisols are soils rich in iron and aluminum oxides and Acrisols or Ultisols are soils in an extreme or ultimate stage of weathering. While the senile soils acquire variable-charge characteristics with age, Andosols (in the FAO system) or Andisols (as proposed for Soil Taxonomy), which form from erupted volcanic glass, have variable charge from birth. In arid regions where leaching intensity is low, variable-charge glass weathers to permanent-charge zeolites or smectites. Andosols, on the other hand, form in humid regions where volcanic glass weathers to hydrated aluminosilicates of short-range order such as allophane and imogolite. An important feature of allophane and imogolite is their high specific surface, which can range from 800 to 1000 $m^2 g^{-1}$. It has been generally assumed that the high organic matter content of Andosols is related to the high specific surface of allophane and imogolite. In fact, the prefix *An* of Andosols and Andisols is a Japanese term for the dark color of organic matter-rich soils formed from volcanic ash. Although we now know that Andosols are not always dark colored, it still holds that they are nearly always high in organic matter content. Andosols are therefore loose, easy-to-work, and physically stable, and they naturally possess the soil qualities organic farmers work so hard to sustain.

It is no accident that some of the most productive, stable, and sustainable agricultural systems are associated with Andosols. One consequence of such richly endowed systems is that they go hand-in-hand with high population density. Rwanda, the most densely populated African country, occurs in such a setting. Indonesia has the world's fifth largest population and yet most of the people are crowded onto the island of Java. Is it mere coincidence that Andosols cover much of Rwanda and Java? We think not. The capacity of these soils to protect and accumulate organic matter may be a key determinant in the genesis of stable and sustainable ecosystems. Is this feature of Andosols purely attributable to their high specific surface? If it is, why do we not find more organic matter in high specific-surface, permanent-charge soils such as Vertisols?

To answer these questions it is necessary to summarize what we know about the origin and characteristics of variable surface charge and the interaction that occurs between soil organic matter and mineral surfaces, as well as the formation and nature of Al- and Fe-humus complexes.

Definition and Origin of Variable Charge in Soils
Permanent Charge vs. Variable Charge

Soil colloids are mixtures of inorganic and organic polymers produced by weathering processes (Sposito and Schindler, 1986). In situations where weathering has not been intense, the inorganic polymers consist mainly of layered aluminosilicates (e.g., smectite), which have electrically charged surfaces caused by structural imperfections in the crystal lattices. This charge is permanent because it is not greatly affected by ambient conditions such as the pH and concentration of electrolyte in solutions in contact with the polymers. If organic matter content is not high, the dominance of these inorganic polymers gives rise to a permanent-charge soil.

In more highly weathered conditions, however, the oxides (hydrous oxides and oxyhydroxides) of iron and aluminum will be in greater abundance. These minerals also exhibit surface electrical charge, caused by protonation and deprotonation of surface hydroxyl groups. The reactions depend on ambient conditions such as pH and ionic strength of soil solution (Figure 1). Clay

Figure 1. Schematic representation of protonation/deprotonation of an iron hydrous oxide surface and resulting surface charge development.

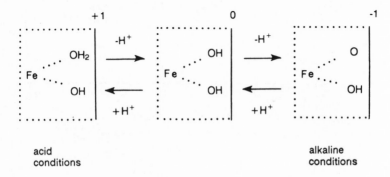

minerals having this type of charge are termed variable-charge clays. By similar protonation/deprotonation reactions on functional groups (carboxylic, phenolic, and enolic) of organic polymers, soil organic matter also contributes to variable charge. The best examples of variable-charge soils are found where the colloidal compartment is dominated by oxidic inorganic polymers and organic matter. Highly weathered soils also contain various amounts of 1:1 clay minerals such as kaolinite and halloysite. These minerals have smaller amounts of permanent charge than does smectite, and also have the variable charge associated with hydroxyl groups at the crystal edges.

Whether or not a mineral soil can be described as a variable-charge soil depends on the mineralogy of the clay fraction and the amount of clay and humus present. Uehara and Gillman (1981) attempted to separate permanent- from variable-charge soils using the proportion of *specific surface* contributing to surface charge (Figure 2). Note that these limits are calculated on the percentage of specific surface and not on a weight basis. Thus a sample containing 95% variable-charge quartz sand and 5% permanent-charge smectite by weight would fall into the permanent-charge category as would a soil containing 80% kaolinite and 20% smectite.

Figure 2. A proposed classification of permanent- and variable-charge soils based on the proportion of specific surface represented by permanent and variable charge, respectively. (Source: Uehara and Gillman, 1981.)

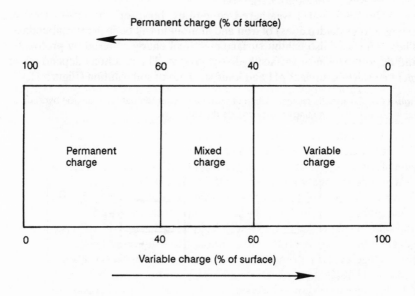

Dynamics of SOM in Tropical Ecosystems

This separation implies that we are able to quantify the permanent and variable charge of a soil; to do this, however, we need operational definitions that will allow experimental procedures to be devised. The following set of definitions has been formulated mainly to be of assistance in soil management and are not as chemically rigorous as some published elsewhere (e.g., Sposito, 1984).

Operational Definitions

Electrical charge on soil particle surfaces can be arbitrarily divided into four categories: negative permanent charge, positive permanent charge, negative variable charge, and positive variable charge. The complex interactions among these charges lead to an adsorption of cation and anion counterions to maintain overall electrical neutrality, giving rise to a cation exchange capacity (CEC) and an anion exchange capacity (AEC). The adsorbed cations and anions are deemed exchangeable if they can be readily displaced by leaching with other electrolyte solutions. The adsorption of such "index ions" is the easiest and most common means of estimating the *effects* of surface charge, but it must be remembered that surface charge itself is not being directly quantified.

In soils containing both variable and permanent charge, values for CEC and AEC can be equal at some particular pH value and this pH can be designated the *point of zero net charge* (PZNC). The determination of PZNC is method dependent, because index cations are adsorbed with various degrees of affinity to the soil surface and their subsequent removal depends on the replacement ion used. Stoop (1980) argued that because Ca^{+2} is usually the dominant exchangeable cation in soil and because of its widespread use as a soil amendment, Ca^{+2} should be the preferred index cation. Monovalent ions such as Cl^{-1}, NO_3^{-1}, or ClO_4^{-1} are often used as the index anion. Clearly, an agreed uniformity of index ions used would be helpful to compare the PZNC of various soils, if this pH value is considered a fundamental soil property.

As shown above, variable charge can be negative or positive, depending on pH, and there will be a particular pH value where net variable charge will be zero — i.e., there will be an equal number of protonated and deprotonated sites. We designate this pH value as the *point of zero variable charge* (pH$_0$). Whereas PZNC is derived from a consideration of whole soil, pH$_0$ refers to the sum of soil variable - charge components only. Since the magnitude of protonation and deprotonation depends on the ionic strength of the ambient solution, increases in electrolyte strength will cause positively charged surfaces to become more positively charged when more protons are adsorbed from solution, and solution pH will rise; conversely, negatively charged sites will become more negative by releasing protons to solution, and solution pH will fall (Figure 3). If a change in ionic strength does not alter the solution pH, there must have been equal numbers of negative and positive variable-charge sites. This immediately suggests a method for determining pH$_0$, but once again there is the

Figure 3. Changes in solution pH when solution electrolyte concentration (n) is increased in the presence of a variable-charge surface. Above pH_0, pH decreases from pH_1 to pH_2; below pH_0, pH increases from pH_1 to pH_2.

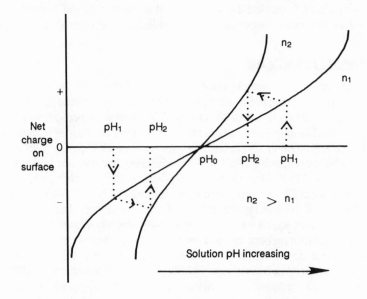

complication that the protonation/deprotonation reactions depend on the actual electrolyte ions used. As argued before, it appears logical to use ions commonly found at the soil-solution interface, e.g., Ca^{+2} and Cl^{-1}. As the procedure suggests, this point of zero charge has also been termed the *point of zero salt effect*, PZSE (Parker et al., 1979). The importance of the position of pH_0 in relation to cation and anion retention will be discussed later in this chapter.

If the pH_0 point is identified, any excess adsorption of exchangeable cations or anions at this pH can be attributed to permanent charge. Only *net* permanent charge would be measured, but in most cases the permanent charge is negative. Tessens and Zauyah (1982) have presented evidence for the existence of permanent positive charge in Malaysian Oxisols. In any event, if the difference between CEC and AEC at pH_0 represents net permanent charge, its subtraction from (CEC - AEC) at any other pH allows an estimation of net variable charge at that pH.

The change in solution pH brought about by increasing the solution electrolyte concentration is usually referred to as delta pH, and has often been measured as the difference in pH measured in water, and in $1M$ KCl. We believe that this difference between electrolyte concentrations is extreme and problems

Dynamics of SOM in Tropical Ecosystems

arise if significant amounts of KCl-extractable Al are present. In keeping with the sentiments expressed above, the difference in solution pH measured in 0.002M CaCl$_2$ and in 0.02M CaCl$_2$ should serve as a more useful definition of delta pH.

Extent of Variable-Charge Soils

If we emphasize the agriculturally important surface horizons, we can consider only a few soils to be purely constant charge soils (because of the presence of variable-charge organic matter) and, similarly, we can identify only a few to be variable-charge soils. As indicated in the previous section, agreed criteria for allocating soils to permanent- or variable-charge classes are not yet available.

It is possible to identify certain types of soil having the greatest likelihood of being variable-charge soils using the classification system Soil Taxonomy (Soil Survey Staff, 1975). The diagnostic criteria used in higher level taxa depend on clay mineralogical composition, as does the origin of variable charge. The relationship between mineralogy and soil order in Soil Taxonomy for mineral soils is depicted in Figure 4. Andosols, and to a lesser extent Oxisols, with their high contents of short-range order or oxidic clay minerals are soils with high amounts of variable charge. The Vertisol order has the best examples of permanent-charge soils, while soils in the orders Spodosols, Ultisols, and possibly Inceptisols will have some soils that *could* fulfill agreed criteria to be classified as variable-charge soils.

Figure 4. The relationship between mineralogy and soil order of Soil Taxonomy (Data from Uehara and Gillman, 1981.)

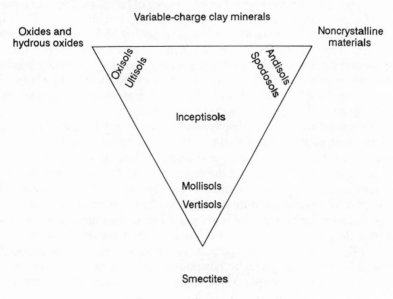

Dudal (1976) estimated that there are about 1 billion ha of Ferralsols (similar to Oxisols) and about 1 million ha of Andepts (Andosols) in the world. Although these represent *in toto* only about 9% of world soils, they are of great agricultural significance because they often occur in climates where water is not a limiting factor to plant growth. It is important, however, that their physical and chemical properties be well understood if optimum and sustainable production is to be obtained from them.

Interaction between Organic Molecules and Oxides

The interaction of small organic molecules with variable-charge surfaces may occur by coulombic or electrostatic attraction, through water bridges, and by ligand exchange. For noncomplexing anions, simple anion exchange reactions apply. Water bridges represent a closer approach of the anion to the charged surface. In ligand exchange, oxygen atoms of dissociated functional groups are coordinately linked to Al or Fe atoms located just beneath the surface layers of ions and water molecules.

For near perfect surfaces of goethite, hydroxyl groups may be one, two, or three coordinated with surface Fe atoms, but only the singly coordinated hydroxyl ions are involved in ligand exchange. Hydroxyls on the 001 surface of gibbsite do not interact with ligands and all the ligand exchange takes place on edge faces (Parfitt et al., 1977a, 1977b). However, more adsorption occurs on gibbsite than on goethite with similar specific surface area.

In soils it seems likely that the mineralogy of the oxide does not have a large influence on the nature of the oxide surfaces because they will all be hydrated, and surface layers will be less than perfect with respect to atomic orderliness. Thus the surface of hematite is probably not different from that of goethite, and the major factor controlling adsorption of organic materials will be the surface area of the oxide. Secondary factors will include the pH as it is influenced by the adsorption of other anions, such as phosphate and silicate, which compete with organic anions for surface sites.

In general the surface complexation of simple organic acids is predictable from work such as that of Parfitt et al. (1977a, 1977b) and Kummert and Stumm (1980). The tendency of organic ligands to form surface complexes on oxide surfaces is similar to that of the organic ligands to form complexes with metal ions in solution. Monoprotic acids such as acetic and benzoic are weakly sorbed; polyprotic acids are more strongly adsorbed. In both cases — weak and strong — adsorption is greater at lower pH values and for monoprotic acids is greatest at the pK_a of the carboxyl group and then decreases at pH values lower than the pK_a (Figure 5). For compounds with functional groups that dissociate at high pH values, e.g., phenolic compounds, adsorption could increase at higher pH values. At a given pH, adsorption will be greatest on the oxide surface

Figure 5. Adsorption envelopes for acids on Al oxides. (Data from Kummert and Stumm, 1980; Parfitt et al, 1977c.)

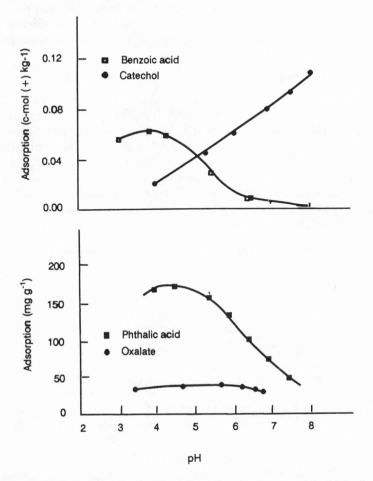

with the highest pH_0. Generally adsorption is decreased in high electrolyte concentration, but this depends on the nature of the cations.

Organic Cations

Organic cations can be "excluded" from the variable-charge surfaces of allophane. Wada (1985) has pointed out that large organic cations, such as long-chain alkylammonium compounds and cetylpyridinium and piperidinium ions, are not retained on the surfaces of allophanic materials. These large molecules are sieved out and cannot penetrate the pores that contain the exchange sites.

The exclusion of cationic organics from oxide surfaces may have implications for the use of cationic herbicides in variable-charge soils. For example, the phytotoxicity of paraquat and diquat in variable-charge soils may remain high, whereas they are inactivated in soils containing smectites. Similarly, problems could arise from the use of the range of s-triazines as preemergent herbicides in variable-charge soils.

Humic Substances

Studies of adsorption of isolated humic substances on synthetic oxides have been tabulated by Oades and Tipping (1988). The pH dependence of adsorption of organic acids on oxides is reflected in the adsorption of humic substances on oxides. In general, adsorption is greater at lower pH values, suggesting that adsorption is controlled by carboxyl groups. Recent work with ^{13}C NMR (nuclear magnetic resonance) of soil organic matter has indicated that carboxyl groups are even more dominant than is indicated by chemical determinations of functional groups. Some adsorption isotherms of fulvic acids on gibbsite, goethite, and imogolite are shown in Figure 6.

Figure 6. Adsorption isotherms and envelopes of fulvic acids on variable-charge minerals. (Data from Parfitt et al, 1977b.)

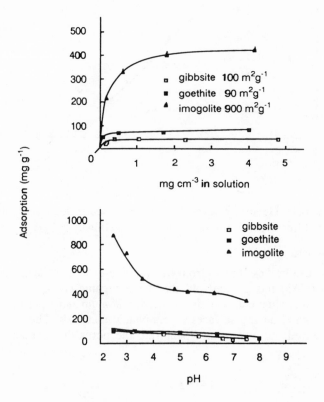

Dynamics of SOM in Tropical Ecosystems

The graphs in Figure 7 show strong adsorption of humic substances up to about 50 mg g^{-1} at pH values of many variable-charge soils. This means that pristine oxide surfaces would adsorb polycarboxylic organic materials from most natural waters and indicates that in soils most oxide surfaces will be covered by organic polyanions (in competition with silicate and some phosphate).

Figure 7. Adsorption isotherms and envelopes of three humic acids from lake waters on geothite. (Data from Tipping, 1981.)

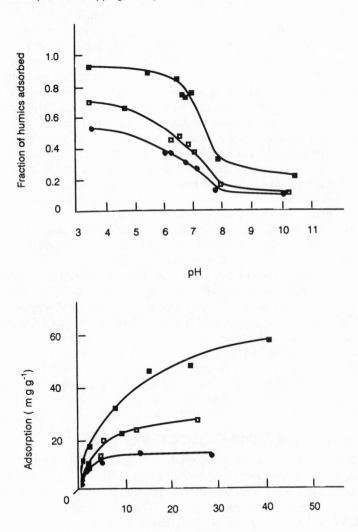

It is more realistic to consider the adsorption on an areal basis rather than a gravimetic basis. Data collected by Oades and Tipping (1988) indicate that the adsorption density is about 1 to 2 mg m^{-2} and as shown by the adsorption envelopes (see Figure 6) is greatest at lower pH values. Because the major interaction is between carboxyl groups on the organic molecule and the oxide surface, adsorption is influenced by the density of carboxyl groups on the organic macromolecules. It is clear from current data that adsorption is influenced by molecular size. Parfitt et al. (1977a) showed that all singly coordinated hydroxyl groups on a goethite (400 mol g^{-1}) could be exchanged by phosphate but only 100 mol g^{-1} by a fulvate preparation. Similarly, fulvate was about one-half as effective as oxalate in exchanging hydroxyl groups on edge faces of gibbsite. Tipping (1981) showed that adsorption expressed on a mass basis increased with molecular weight, but not if adsorption was expressed on a molar basis. For adsorption of macromolecules, steric factors control the amount of adsorbant that can approach the surface and also the fraction of carboxylate groups that can interact with the oxide surface. Thus, when adsorbed, polyanions such as humic substances will create a new variable-charge surface, but the charge will be dominantly negative and the pH$_0$ will be low.

The presence of Ca and Mg ions generally enhances adsorption of humic substances on oxides, presumably by influencing the properties of the adsorbed species, but the actual mechanisms remain undefined.

While the dominant interaction of humic substances with oxide surfaces is one of ligand exchange, where protons are consumed or hydroxyls are released, non-ionic organic materials, e.g., undissociated fulvic acids (Parfitt et al., 1977c) and polyvinyl alcohol (Kavanagh et al., 1976) are also adsorbed. Other specific information on this subject is lacking.

One specific group of soils with variable charge is the Andosols. According to Wada (1985) only Andosols containing allophane or imogolite with an SiO$_2$/Al$_2$O$_3$ ratio close to 1 carry similar amounts of positive and negative charge at field pH values. It was suggested that adsorption of anionic materials on these minerals would be similar to gibbsite. Parfitt et al. (1977c) showed that little carboxylate was formed when fulvic acid interacted with imogolite; however, adsorption with imogolite was almost tenfold more than adsorption with gibbsite, due to a surface area approaching 1000 m^2 g^{-1}. Laboratory work on adsorption of polyanions on allophane or imogolite appears to be lacking, but further information comes from studies of natural organo-oxide complexes.

Natural Hydrous Oxide-Organic Complexes

Natural associations of organic materials with variable-charge minerals have been examined by extractive procedures. The interaction of organic polyanions with hydrous oxides inevitably reduces the positive charge on the oxide surface and the pH$_0$ of the organo-oxide complex is low. Removal of organic matter should have the reverse effect. This has been confirmed several times

(Tweneboah et al., 1967; Moshi et al., 1974; Sequi et al., 1980; Cavallaro and McBride, 1984), although it is not clear whether the positive sites produced represent sites exposed on an existing oxide surface or new sites created by hydrolysis of Al or Fe released during destruction of organic matter. Moshi et al. (1974) presented a negative linear correlation between positive charge and organic matter content for a tropical red loam. It would seem that loss of organic matter due to cultivation will expose positive sites, which will increase anion and ligand exchange, e.g., phosphate sorption, and change the pH_0 and the physical properties of the soil. However, an increase of phosphate sorption with increasing organic C content has been described for Andosols by Mizota et al. (1982).

A range of chemical procedures has been used to extract iron oxides from various parts of the continuum that exist from crystalline oxides, through finely divided goethites and short-range order minerals such as ferrihydrite, to organic complexes of Fe and Al (Figure 8).

Figure 8. Chemical extraction of Fe oxides and complexes from soils.

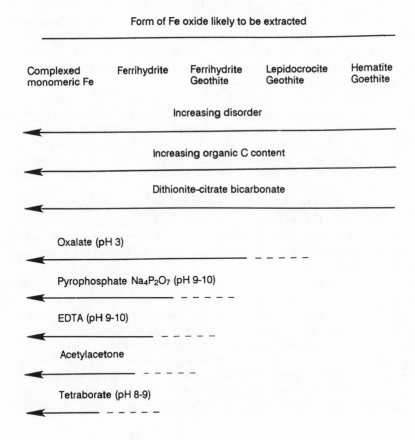

Form of Fe oxide likely to be extracted

| Complexed monomeric Fe | Ferrihydrite | Ferrihydrite Geothite | Lepidocrocite Geothite | Hematite Goethite |

Increasing disorder

Increasing organic C content

Dithionite-citrate bicarbonate

Oxalate (pH 3)

Pyrophosphate $Na_4P_2O_7$ (pH 9-10)

EDTA (pH 9-10)

Acetylacetone

Tetraborate (pH 8-9)

The problem with the various reagents is that they are not specific, but the first three (dithionite-citrate biocarbanate, oxalate, and pyrophosphate) have proved useful in differentiating crystalline oxides from ferrihydrite, while pyrophosphate and EDTA supposedly extract "organic Fe." In fact, both complexing agents peptise and disperse oxide particles a few nm in diameter, which have been shown to be goethite and ferrihydrite. These particles were dispersed because they were coated with organic anions.

As might be anticipated, the crystalline oxides with low surface areas are not quantitatively important in interactions with organic materials. Such oxides occur in soils formed in hot climates with at least one dry period during the year. Linear regressions between C and Al and Fe in oxalate and pyrophosphate extracts have been obtained for a number of variable-charge soils, showing that there is a direct relationship between ferrihydrite or very fine oxide particles and organic matter contents. In fact, the relationships are as expected from the model systems described earlier (Wada and Higashi, 1976; Adams and Kassim, 1984; Evans and Wilson, 1985). Evidence regarding the nature of the finely divided iron oxides was accumulated by Oades (1988a) and is based on C/(Al + Fe) or C/Fe ratios in extracts, which indicate that the Fe is present as small hydroxy-iron cores. In some cases these hydroxy-iron cores have defied positive mineral identification; occasionally goethite and ferrihydrite have been identified and in the absence of X-ray diffraction, Mossbauer spectroscopy has indicated numbers of adjacent Fe atoms.

One is led to the conclusion that the hydrous iron oxides associated with organic matter are spheres up to 5 nm in diameter that can be dispersed as negatively charged colloids when coated with organic polyanions. When sufficient organic matter is present, the oxide particles are bound together in a matrix of organic matter and the specific surface area of ferrihydrite is inversely related to C content. Treatment with H_2O_2 increased the surface area by more than $100 \ m^2 \ g^{-1}$ (Schwertmann et al., 1986).

The occurrence of monomeric Fe^{+3} complexed by organic matter in soils is equivocal. The hydrolytic tendencies of Fe^{+3} are such that little Fe is likely to exist in this form and would have to be present in stable chelate complexes with molecules such as siderophores (e.g., Waid, 1975; Cline et al., 1983).

The Al-organic system differs considerably from that of Fe. There does not appear to be a continuum from crystalline oxides such as gibbsite to monomeric Al^{+3}. Gibbsite will form in the absence of complexing polyanions and is crystalline. There does not appear to be an Al oxide system with short-range order such as ferrihydrite, possibly because the Al interacts with Si to form disordered aluminosilicates including allophanic materials. However, it is clear that considerable amounts of Al exist as monomeric species that interact electrostatically with carboxyl groups on organic materials. This seems to apply to most variable-charge soils, e.g., Andosols and B horizons of

Spodosols. This is probably because Al^{+3} is two orders of magnitude more soluble than Fe^{+3}. Thus $[Al^{+3}]$ in solution exceeds $[Fe^{+3}]$ by a factor of 10 or more. For soil systems there is evidence that much of the positive charge in variable charge soils is due to Al surface ions rather than Fe because Al is more mobile. This would not be the case in anaerobic situations.

Adsorbed Organic Materials

The nature of the adsorbed organic species remains controversial, but there are indications that the organic materials are dominated by humic material with relatively low molecular weights, i.e., not more than a few thousand Daltons. This means that a large proportion of the organic matter is extractable using acids or complexing agents such as pyrophosphate. These statements are best illustrated by work on B horizons of Spodosols and by the characteristics of humic materials in Andosols with humic/fulvic ratios between 1 and 2 (Wada, 1985). In our experience the organic materials brought into solution by treatments with sodium dithionite are aliphatic, with significant carbohydrate contents and carboxyl groups. This information was obtained on fractions greater than 1000 Daltons using ^{13}C NMR. The spectra in Figure 9 were obtained from the ferrohumic horizon of a Queensland Spodosol, a ferrihydrite from the Northern Territory of Australia, and a Rhodoxeralf. Based on the literature, we had assumed that aromatic acids might have accumulated on oxide surfaces,

Figure 9. ^{13}C NMR Spectra of dithionite and ammonium oxalate extracts of soils and oxide-rich materials. Samples were desalted by ultrafiltration and represents molecules greater than 1000 daltons.

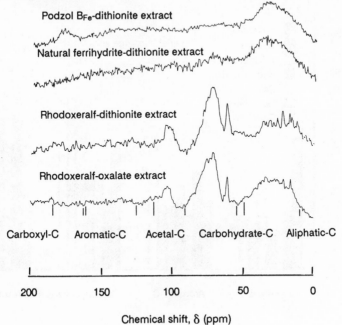

but in all situations examined so far aromatic-C has not been detected in dithionite or oxalate extracts by NMR spectroscopy. It is quite clear that polymethylene (aliphatic) components and carbohydrates are dominant.

Composition of Organic Matter in Variable-Charge Soils

K.R. Tate and Theng (1980) suggested that the composition of organic matter in variable-charge soils is not different from the composition of organic matter in other soils. However, recent unpublished work involving solid state ^{13}C NMR (Figure 10) indicates that some soils with variable charge are richer in aliphatic materials and carboxyl groups than are other soils. There are indications that Andosols have higher ratios of hexoses to pentoses than has been recorded for other soils. This would indicate a higher proportion of microbial polysaccharides in Andosols than in other soils (Oades, 1972; Murayama, 1988), stabilized presumably by adsorption to variable-charge surfaces and trapped in fine pores.

There are indications that extensive oxide surfaces lead to higher organic matter contents, but the question remains controversial because of different environmental factors. There are laboratory data, however, that show stabilization of organic materials by clay systems with variable charge. Mineralization of C in variable-charge allophanic soils has been shown to be considerably less

Figure 10. Distribution of carbon in chemical groups as shown by solid state ^{13}C NMR of seven soils (Data from Abboud and Turchenek, 1986; Preston et al, 1987; Oades et al 1988.)

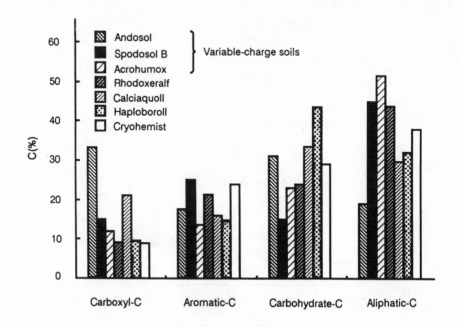

Dynamics of SOM in Tropical Ecosystems

than that in other soils (Figure 11). While various other factors in these soils were not constant, e.g., clay content and soil structure, particularly at the microscale, the differences are large enough to indicate a considerable stabilization of C in the allophanic soils (Martin et al., 1982). These differences were confirmed when various ^{14}C-labeled compounds were added to the soils and allowed to mineralize. Similarly, addition of allophane to soils decreased the mineralization of various ^{14}C-labeled substrates (Zunino et al., 1982).

Part of the stabilization of C compounds in variable-charge soils is no doubt due to the adsorption of polycarboxylic materials on oxide surfaces by mechanisms already discussed, but another factor concerns the microstructure of the organo-variable-charge-clay systems. This can be well illustrated by the mineral imogolite, which consists of bundles of hollow fibers. To associate with the surface of imogolite, organic molecules must be small, as large molecules would be excluded from such a finely porous system. The small molecules are then protected from enzymes and organisms. Andosols are characterized by small pores capable of trapping organic materials, which are not then accessible to organisms or enzymes. The considerable microporosity of such soils is created by the hydrous oxide-humus interaction, and it is this microstructure that stabilizes organic C to the extent that mean residence times for C in allophanic soils are in the range of several thousand years (Wada and Aomine,

Figure 11. Mineralization of carbon from soils with permanent- and variable-charge colloids. (Data from Martin et al, 1982.)

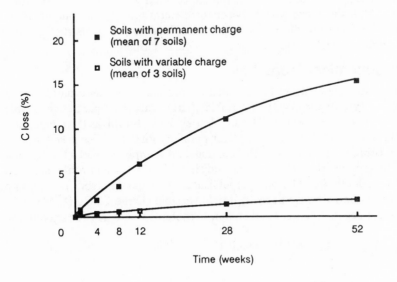

1973). Similar situations can occur in other soils dominated by fine-grained hydrous oxides, e.g., Oxisols. The stability of the microstructure and clay dispersion depends on the PZNC, which is controlled by the relative quantities of oxide surfaces and organic matter. Most variable-charge soils seem to be characterized by strong aggregation of colloidal material into silt-sized particles. Thus, there is a role for surfaces, for very small aggregates, and for large aggregates in the stabilization of organic matter in Andosols. However, data tabulated by Wada (1985, 1986) show that with cultivation the structural system deteriorates and organic matter contents fall significantly. There is no means of stabilizing organic materials completely in soils subjected to regular cultivation.

Interactions Between Organic Matter and Variable-Charge Clays

Although we can describe in some detail, or speculate about, some of the chemical reactions that may take place between the various clay minerals and organic compounds known to exist in soil, the reality is that these entities cover a range of chemical states and combinations. Silicate minerals are often coated to various degrees with hydrous oxides and oxyhydroxides, and "organic matter" is present in all stages of decomposition—from freshly incorporated plant material, through living and dead microbial cells, to the decomposed remains of both plant and animal tissues. The task, therefore, of identifying specifically the reactions on-going in a particular soil is a daunting if not impossible one. Nevertheless, we *are* able to study the effects on a variety of soil properties, resulting from the interactions between the mineral and organic components in a soil. The following discussion will be confined to the properties of soils referred to as variable-charge soils as previously defined.

Electrochemical Properties

There is sufficient evidence to show that the presence of organic matter causes the cation exchange capacity of variable-charge soils to be greater than what would pertain in its absence. The simplest evidence for this is the (sometimes rapid) decline in CEC with soil depth, in parallel with the decline in soil organic matter content, even though the content and composition of clay has not changed appreciably (Table 1). In a study involving 26 Oxisol profiles sampled to a 90-cm depth, G.P. Gillman (unpublished data) found a strong relationship between cation exchange capacity due to variable charge CEC_V (total CEC - permanent negative charge) and organic C content. The regression equation is:

$$CEC_V = 1.32 + 1.09 \text{ organic C}, \ r^2 = 0.76$$

Table 1. Data for a north Queensland (Australia) Oxisol under virgin rainforest showing a decline in CEC and organic carbon with depth, but relative constancy of clay content and soil pH.

Depth (cm)	Organic carbon (%)	CEC (cmol(+) kg^{-1})	Clay (%)	Soil pH
0 - 15	3.9	3.2	74	5.0
15 - 30	2.2	1.0	67	4.8
30 - 60	1.4	0.5	66	4.7
60 - 90	0.7	0.3	69	4.7

In other words, a 1% increase in organic C causes CEC to increase by one unit. Other workers (e.g., Tinker and Ziboh, 1959; Hawkins and Brunt, 1965) have recorded greater increases in CEC per unit of percentage of organic C, but values are confounded by the method used for determining CEC.

One reason for this phenomenon lies in the effect of organic matter on the point of zero charge of the soil variable-charge components (pH$_0$). Figure 12 illustrates the relationship between point of zero charge of variable-charge components and organic C for 24 samples from six Oxisols occurring under virgin rainforest in north Queensland. Increasing levels of organic C result in a rapid lowering of the zero point from about pH 6.5 to below pH 4.0.

The greater the difference between soil pH and pH$_0$, the greater the net surface charge on variable-charge components, and if (pH$_0$ - pH) is less than zero, this net charge is negative. For the 24 samples illustrated in Figure 12, (pH$_0$ - pH) was negative for ten samples, and the CEC attributed to variable charge was found to increase sharply with decreasing pH$_0$, or with increasing negativity of (pH$_0$-pH), as shown in Figure 13. As the *permanent* charge CEC of these soils is generally only 1 or 2 cmol(+) kg^{-1}, the importance of maximizing the CEC due to *variable* charge is obvious. Maintaining organic matter contents at the highest possible levels to achieve minimum pH$_0$ values, combined with amendment practices that keep soil pH values at the highest practical levels, will contribute greatly to the capacity of these soils to retain plant nutrient cations.

The above-mentioned relationships observed in Oxisols should also be found in the extreme examples of variable-charge soils, viz. the Andosols, in which the mineral composition is dominated by allophane, imogolite, and noncrystalline hydrous oxides. In addition, organic matter contents are often unusually high as a result of strong chemical bonding between the mineral clays or polymer hydroxyions and humic substances (Wada, 1985). Pardo and Guadalix (1988) examined eight Andosol samples (which had been air-dried) from Spain and the Canary Islands and found no clear correlation between pH$_0$

Figure 12. Relationship between pH_0 and organic carbon content in 24 samples from six Oxisols under virgin rainforest in northern Queensland, Australia.

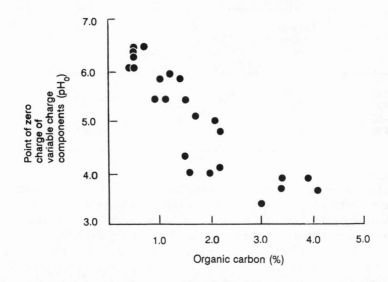

and organic matter content. However, Radcliffe and Gillman (1985) found a very strong negative correlation between organic C content and pH_0 for 12 samples from Andosols formed on airfall tephra in Papua, New Guinea, and a similar relationship has been reported by Sollins et al. (1988) for Andosols formed on alluvially resorted ash in Costa Rica.

There are at least two likely mechanisms that could explain the depression of the soil pH_0 value by organic matter. First, organic matter itself has a low overall point of zero charge, caused by a preponderance of carboxyl groups that deprotonate at low pH values. Since the soil pH_0 value represents the mean of all of the protonated and deprotonated sites on all of the variable-charge components, an increase in deprotonated humic material would result in a lower overall pH_0 value. Second, reaction between negatively charged organic anions and positively charged oxide sites would favor a dominance of deprotonated (negatively charged) sites, i.e., a lowering of the overall pH_0. It would be worthwhile to elucidate the mechanism by which organic matter depresses the pH_0 value in variable-charge systems, because it is generally true that systems can be more easily manipulated or managed if their intrinsic nature is well understood.

It follows from the above discussion that the lack of organic matter in the subsoils of oxidic soils will result in high pH_0 values and an appreciable soil anion exchange capacity. We need to know if advantage can be taken of these

Dynamics of SOM in Tropical Ecosystems

Figure 13. Increase in CEC of variable-charge components with increasing negativity of (pH$_0$-pH) in northern Queensland (Australia) Oxisols (from Figure 12).

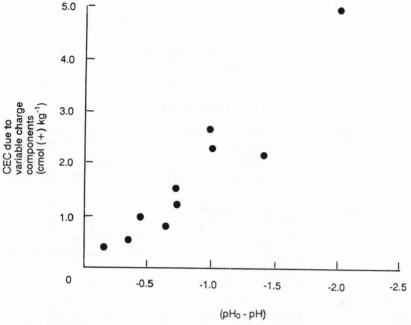

separate situations occurring in a single profile to allow us to manage the retention of both cations in the surface horizons and anions in the subsoil. This may be especially important in disturbed systems when NH_4^+-N is converted to NO_3^- N in the surface horizons. Subsequent leaching into the subsoil could result in storage on positively charged sites, where it could be available to deep-rooted plants.

Soil pH and Soil Acidity

Although processes such as the amount of CO_2 dissolving in the soil water and the forming of organic acids by microbial decomposition influence soil acidity, the exchange reactions between soil solution and the soil particle surfaces are the main regulators of soil pH. The relative amounts of basic and acidic exchangeable cations determine its actual value.

The inorganic components of variable-charge soils, particularly the Fe and Al oxides, allophane, and imogolite, are weakly acidic when compared with permanent-charge minerals such as smectite. In the latter, structural defects in crystal lattices cause a deep-seated and clear separation of negative and positive charge. Thus in base-depleted situations, exchangeable Al is more easily dissociated and the clay acts as a strong acid. In the former, Al is more intimately associated with surface functional groups and is not as readily dissociated.

Moreover, the oxides have a high point of zero charge, and soil pH tends to drift toward these higher values leading to hydrolysis or polymerization of any exchangeable Al.

The addition of organic residues to variable-charge soils and subsequent mineralization and nitrification of N could result in increased acidity and solubilization of Al from mineral sources. Additions of lime or dolomite with the residue would create a favorable environment for controlling residue decomposition as well as for maintaining a steady base-saturated CEC.

Adsorption of Phosphate and Sulfate

The clay mineralogical composition responsible for variable charge in soils often causes these soils to have a large capacity to adsorb or absorb polyvalent anions such as phosphate, sulfate, molybdate, selenite, and the like. Here we consider only the two agronomically important species, phosphate and sulfate.

Many aspects of the sorption of phosphate by variable-charge soils have been studied by R. L. Fox (University of Hawaii) and co-workers. These studies have highlighted the importance of the mineralogical composition of the clay fraction in determining the quantities of P sorbed, with Andosols, followed by Oxisols, achieving the highest P-sorption capacities.

The presence of organic matter ameliorates the amount of P sorbed by Oxisols. Fox and Searle (1978) point out that this might be the direct result of competition between organic anions and phosphate for binding sites on iron and aluminum oxides or hydrous oxides, and also that by lowering the point of zero charge (pH_0), there are fewer positively charged sites for attraction of phosphate ions to the vicinity of the soil surface. In comparing forested and cultivated Oxisols in Kenya, Moshi et al. (1974) found a P-sorption maximum of 560 mg P kg^{-1} soil for the forested soils (6.8% C) compared to 800 mg P kg^{-1} soil for the cultivated soil (3.8% C). Figure 14 summarizes the results for 14 Oxisol profiles from north Queensland where P-sorption capacity was plotted against citrate-dithionite extractable iron and aluminum content. The samples from the organic-rich 0-15 cm layer sorbed much less phosphate than the subsurface soils at any particular sesquioxide content.

The above results described suggest that when virgin Oxisols are cultivated, the resulting decrease in organic matter content will cause P-sorption capacity to increase. This will happen if little or no phosphatic fertilizer is applied, but there are reported instances where high and continuous applications of superphosphate has actually caused a decrease in P sorption relative to the virgin state. For instance, Toreu et al. (1988) compared undisturbed Oxisols (3.9% C) with cultivated and well-fertilized Oxisols (1.6% C) in north Queensland and found P-sorption capacities of 555 and 494 mg P kg^{-1} soils, respectively.

Figure 14. Relationship between P-sorption capacity and sesquioxide content for 14 Oxisols formed on basalt in north Queensland, Australia.

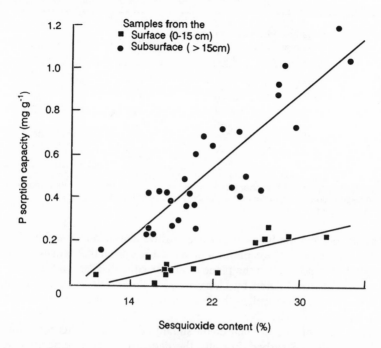

In very high P-sorbing Andosols, the correlation between P sorbed and organic matter is *positive* (Wada, 1985). The P-sorption capacity of Andosols rich in Al- and Fe-humus complexes is more highly dependent upon solution P concentration than that of Andosols where allophane and imogolite are the principal constituents. In other words, the association of organic matter with Fe and Al causes phosphate to be more specifically adsorbed than when the Al is present in these secondary minerals. In the latter case, the sorption mechanism is more electrostatic.

After examining the results of several workers who explored the relationships between organic matter and phosphate in Andosols from Japan, New Zealand, and Colombia, Fox (1980) concluded that organic matter is most stable in acid, P-deficient, allophane-rich soils. This infers that large additions of phosphate to such soils could result in increased organic matter decomposition and the effect of this on other properties such as aggregate stability would need to be carefully monitored.

It is clear that sulfate is not as strongly sorbed by variable-charge soils as is phosphate, although there is a specificity compared to monovalent ions such as chloride and nitrate. Marsh et al. (1987) studied a number of New Zealand pasture soils, many of which were formed on volcanic ash, and found

Table 2. The effect of organic matter content and sign of net charge on the extractable sulfate content of a virgin Oxisol from high rainfall coastal Queensland, Australia.

Depth (cm)	Organic matter (%)	Phosphate extractable sulfate $(\mu\ g^{-1})$	delta pH[a]	Clay (%)
0 - 10	11.5	34	- 0.90	60
10 - 20	6.8	69	- 0.92	80
20 - 30	4.6	58	- 0.39	86
30 - 60	3.1	240	- 0.10	78
60 - 90	2.0	320	- 0.04	nd
90 - 120	1.3	370	+ 0.15	75
210 - 240	nd	630	+ 0.50	65

[a] delta pH = (pH 1NKCl - pH H_2O). The sign of delta pH indicates net charge of soil variable-charge components if exchangeable Al is low. Methods for determining delta pH have since been modified. See section on Operational Definition for explanation. nd = not determined.

a very strong correlation between surface positive charge and S sorbed. They concluded that the sulfate ion enters a plane of adsorption that is distinct from the surface but is closer than the plane of nonspecifically adsorbed ions. Such a mechanism was postulated by Uehara and Gillman (1981) to explain the increase in solution pH as well as the increase in the pH_0 when sulfate is added to an oxidic soil.

The effect of organic matter on sulfate adsorbed by this mechanism would be to reduce S sorbed, because the decrease in pH_0 associated with increased organic matter content would cause a reduction in surface positive charge. Thus Couto et al. (1979) observed that surface horizons of two Brazilian Oxisols sorbed less sulfate than their corresponding subsurface horizons. Amounts of sulfate extracted from various depths in an Oxisol occurring under virgin rainforest in coastal tropical Queensland, along with some relevant soil properties, are summarized in Table 2. It appears that inputs of sulfur from rainfall are not retained by the organic matter-rich surface horizons but are stored in the net positively charged subsoil. Thus, provided that root proliferation in the subsoil occurs, which requires among other conditions sufficient Ca and Mg, these subsoils can act as an important reservoir for sulfate sulfur.

Physical Properties

One of the main benefits arising from the addition of organic materials to soil is an improvement in soil physical properties, in particular the soil structure, brought about by the stabilization of soil aggregates by the products of microbial activity. The maintaining of granular structure, and therefore of free internal drainage, is particularly important in variable-charge soils, which often occur in regions subject to intense rainfall, as they would be prone to erosion of the topsoil under high runoff conditions.

There are inherent difficulties in identifying the actual organic species responsible for aggregate binding, but a good deal of evidence (reviewed by Lynch and Bragg, 1985) points to the involvement of soil polysaccharides. These large molecules contain many hydroxyl and carboxyl functional groups that would be deprotonated and hence negatively charged at soil pH values. As discussed earlier, we could therefore expect strong association with positively charged sites on oxides leading to well-developed microstructure. This type of mechanism would explain why no water-dispersible clay was found in an Oxisol at such a depth in the profile that an appropriate balance of oxides and organic matter caused the net surface charge to be zero (Gillman, 1974) (Figure 15). Above this layer, an accumulation of organic matter resulted in net negative surface charge and hence an appreciable water-dispersible clay content, while below the layer, excess positive charge again caused clay to disperse in water.

Figure 15. Distribution of water-dispersible clay content in a north Queensland (Australia) Oxisol. For information on profile clay and organic matter content, see Table 2.

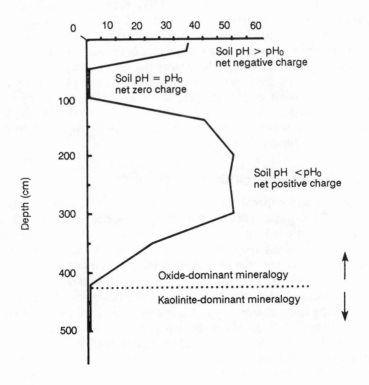

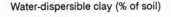

Polysaccharides themselves act as important food sources for many types of bacteria, inferring that aggregate stability resulting from this mechanism would be short lived (Tisdall and Oades, 1982). However, complexes of polysaccharides with metal cations are resistant to breakdown (J.P. Martin et al., 1966), and the abundance of Fe and Al in variable-charge soils would favor such complexes. Certain humic substances, particularly phenolic compounds, also have the ability to protect polysaccharides from decomposition (Griffith and Burns, 1972).

The effect on aggregate stability of the association between organic matter and variable-charge clay is most striking in Andosols, where organic matter contents of 15 to 20% are typical. These soils have low bulk density, due to high organic matter content and to low particle density. Pore volumes are very high, and although this is partly due to the inherently large pore spaces of minerals such as allophane and imogolite, pore volumes are also very high in Andosols not containing these minerals but having high humus content. Wada (1985) therefore infers that organic matter plays a very important role in inter-particle bonding, leading to the formation of stable aggregates, with the involvement also of "active Al," which is the Al extracted by pyrophosphate, citrate-dithionite, or oxalate. The results of Yasuhara and Takenaka (1977), in comparing surface soil and subsoil of an Andosol before and after compaction, showed that hydraulic conductivity was not affected in the surface horizons after compaction, whereas conductivity was reduced in the subsoil. They postulated that organic matter stabilized voids between the aggregates.

Although it is clear that surface electrical charge plays a key role in determining a number of physical properties of variable-charge soils, more work needs to be done to relate surface charge characteristics to soil physical properties in these soils and to understand the obviously critical importance of organic matter in these relationships.

Conclusions and Recommendations

When considering the specific properties of variable-charge soils and especially the influence of organic matter on these properties, one should keep in mind the variability that can be encountered over extremely small distances, which may not be recognized when bulk soil is subjected to laboratory analysis. Microsite variability will have an important influence on factors such as plant root response and distribution and on soil-solution chemistry.

It is also important that the unique properties of variable-charge surfaces be taken into account in the laboratory analysis of soils that possess significant variable charge. Thus air-drying an Andosol sample can cause irreversible changes to surface area and characteristics. The determination of important soil properties such as cation exchange capacity should be carried out at a pH and electrolyte concentration similar to that existing in the field.

Ultimately we would like to see the development of models that will allow us to make useful predictions about the turnover rate and stability of soil organic matter in soils generally, but such models will need to accommodate the specific properties of variable-charge soils. The soil organic matter model developed by Parton et al. (1987) overestimates organic C for fine-textured soils. Since most Andosols are fine-textured and contain more organic matter than do other soils of equivalent textures, the Parton et al. model may compute the right C values for the wrong reasons.

We need models because the temptation to manipulate soil organic matter will increase as our desire to intervene and conserve natural resources competes with the demands of expanding populations. We lack adequate understanding of processes that govern the fate of soil organic matter so that research to understand processes will continue to receive high priority. But understanding processes is only the first step. Researchers have the added responsibility to organize and condense what we already know to develop dynamic, process models that can predict outcomes of natural and anthropogenic soil disturbances. That we are unable to make useful predictions about the turnover rate and stability of soil organic matter should not deter us but should be the basis for identifying knowledge gaps.

Theme Imperative: In the final analysis, the benefits of model building rest not in the ultimate creation of a reliable user-oriented product but also in the certainty that the exercise will expose large areas of ignorance and help researchers identify profitable areas for research. In this respect, we would suggest that priority be given to specific research issues to assist both model development and our general understanding of the role of organic matter in variable-charge soils.

Research Imperatives:

1. Identify the mechanisms that maintain contents of organic matter in variable-charge soils.

2. Characterize the organic matter pools that affect variable charge in soils, including both specific chemical pools and the functional biological pools used in modeling efforts.

3. Quantify relationships between soil physical properties and surface charge characteristics in variable-charge soils.

4. Define the role of organic matter as a regulator of soil acidity and phytotoxicity in variable-charge soils.

5. Quantify changes in variable-charge properties that can be brought about by clearing and management practices.

6. Develop management strategies consistent with the specific properties of variable-charge soils; especially with P-sorption capacities and the proportions of negative and positive charge throughout the soil profile.

Chapter 4

Biological Processes Regulating Organic Matter Dynamics in Tropical Soils

Jonathan M. Anderson and Patrick W. Flanagan

with Edward Caswell, David C. Coleman, Elvira Cuevas, Diana W. Freckman, Julia Allen Jones, Patrick Lavelle, and Peter Vitousek

Abstract
The processes of litter decomposition and formation and mineralization of soil organic matter are determined by climate, soil conditions, resource quality, and the activities of plant roots, microorganisms, and animals. These variables are generally seen to operate in a hierarchical manner, but examples are considered where biological processes interacting with soil conditions can override the effects of climate on litter decomposition on both local and system-level scales. Resource quality and edaphic conditions are key parameters that influence SOM stabilization by microorganisms in tropical soils. Termites and earthworms contribute to SOM dynamics both directly (by influencing metabolic processes) and indirectly (by affecting litter inputs and decomposition, by forming and disrupting soil aggregates, and by changing soil physical controls on microbial processes). Plant roots also have direct and indirect effects on SOM formation and turnover: roots are important inputs of organic matter to soils, root exudates prime SOM mineralization, and roots affect soil conditions by forming sinks for water, nutrients, and oxygen.

The effects on SOM of temporal and spatial changes in biological processes are considered. These are generally most marked when natural systems are converted to grassland or arable agriculture, but in some cases large pools of stabilized SOM formed under forest cover show little response to the effects of changes in litter inputs, soil conditions, and biota.

The processes of decomposition and SOM formation are essentially similar in temperate and tropical systems. Because of different combinations of variables, higher level expressions of their effects, or longer periods over which processes can produce measurable changes in SOM, studies on tropical soil biological processes can contribute to the general understanding of SOM dynamics in natural and managed systems.

Processes of Decomposition, SOM Formation, and Stabilization

Dead organic matter in soils can be divided into cellular material, including intact litter and particulate organic matter down to about $250\,\mu$m, and noncellular "humus" compounds which make up as much as 98% of the total soil C in most soils (Theng et al., Chapter 1, this publication). The decomposition rate of the litter fraction is much faster than that of humus because a small quantity of organic C in each age-class of litter becomes more recalcitrant with time than of the original plant constituents. These stabilized fractions accumulate until some steady state is reached between C inputs and mineralization, assuming conditions for decomposition, SOM formation, and mineralization remain more or less constant.

Litter decomposition represents the sum of mass losses brought about by catabolism (action of microbial and animal enzymes), comminution (physical breakdown by animals and abiotic processes), and leaching of water-soluble materials (Swift et al., 1979). As a consequence of the change in state brought about by these processes, soluble or particulate materials may be transported down the profile by invertebrates, water, or gravity to sites where different chemical and biological conditions operate different from those regulating decomposition in the parent material. The solubilization of litter constituents and the relocation and precipitation by condensation reactions in mineral soil are important processes in the formation of stabilized humus constituents (Duchaufour, 1977).

The cellular material decomposes rapidly in tropical ecosystems, usually within a few months to a year or two depending upon chemical composition, but during the process of decomposition microbial cell walls and some metabolic products are synthesized, which decompose at least 10 times more slowly than the original litter. About 10-20% of aboveground litter and 20-50% of root litter may be converted to humus (Nye and Greenland, 1960) and the rest is mineralized as CO_2. Within the soil, some fractions of this organic matter undergo further biochemical stabilization and may also be complexed with clays or other minerals or they may be physically protected within clay microaggregates (Edwards and Bremner, 1967). These biochemical and physical processes further increase the resistance of humus compounds to enzyme attack, so that mineralization rates of some fractions may be less than 0.1-1% per year.

Either chemical or physical fractionation of SOM, or a combination of both methods, delimits suites of compounds with radiocarbon ages ranging from modern to several thousand years (Paul and van Veen, 1979). In temperate soils, up to 50% of SOM has a residence time of 1000 years or more and has an insignificant role in the short-term cycling of C and N (Jenkinson and Rayner, 1977). As a consequence of the highly stabilized nature of these fractions, turnover times may be similar in temperate and tropical soils despite the differences in temperature regimes (Sauerbeck and Gonzalez, 1977). Scharpenseel and Schiffman (1977) showed that mean residence time increases with depth for a large number of temperate and tropical soils with material at 2 meters depth in the order of 10,000 years old. Thus there are also gradients of physical and chemical conditions in soil that impose limits on the mineralization rates of these slowly decomposing SOM fractions and which are quite distinct from the decomposition of root and leaf litter in soil at these depths.

Microorganisms usually carry out 80-99% of C mineralization in temperate and tropical soils (Phillipson, 1973; Luxton, 1982), but the soil invertebrates (fauna) can have significant effects on C fluxes by indirectly affecting plant production and microbial processes through their feeding and burrowing activities.

A Hierarchy of Variables Determining Organic Matter Dynamics

The processes of decomposition and SOM formation are primarily controlled by the biodegradability (resource quality) of organic molecules subject to catalysis by enzymes, mainly of microbial origin. Temperature and edaphic factors (oxygen, water, cations, and clay minerals) operate as immediate controls over these processes, and labile C sources are needed to prime the catalysis of most organic polymers and aromatic compounds. At the organism level, porosity, soil structure, water, and temperature interact to control gaseous diffusion and the activities of microorganisms, invertebrates, and plant roots. Plant root systems provide litter, provide inputs of labile C that can prime microbial processes, act as sinks for water, nutrients, and oxygen, and alter soil structure. The larger soil fauna also indirectly affect these physicochemical controls over microbial processes by their feeding and burrowing activities. As the level of study increases through the population to ecosystem level and as higher levels are involved in investigating global C fluxes, these fine-scale interactions are difficult to quantify and it is practical to relate the turnover rates of gross organic matter pools to macroclimate and soil type. Macroclimate, soil physicochemical conditions, the quality of organic matter inputs to soils, and the activities of invertebrates and microorganisms can thus be seen to operate in a hierarchical manner on decomposition and SOM formation (Table 1) (J. M. Anderson and Swift, 1983). Climate sets the gross constraints for rates for litter decomposition and SOM turnover. Soil physicochemical conditions modify, and may even override, climatic controls on belowground litter decomposition and the formation and mineralization of stabilized SOM, so that there may be no direct relationship between surface litter decomposition rates and SOM accumulation. These variables interact and feed back to higher level controls: invertebrates and plant roots can change soil physical conditions influencing SOM stabilization, and feedbacks through soil conditions can change the quality, quantity, and location of litter inputs.

The use of simulation models for SOM dynamics (McGill et al., 1981; Pastor and Post, 1986; Parton et al., 1987) provides a basis for testing the importance of these biological variables on the formation and turnover of SOM fractions. CENTURY (Parton et al., 1987), thoroughly discussed in Chapter 6, is driven by three physical site variables (temperature, moisture, and soil texture) with inputs of plant metabolic-C and structural-C to active, slow, and passive (highly stabilized) SOM pools. Parton et al. (1987) acknowledge that the model is sensitive to historical patterns of land use such as grazing (and site history of forested areas), which affect the quantity and quality of litter inputs to the soil and hence the assumptions for steady-state values for SOM fractions. Data for tropical soils rarely differentiate C pools other than litter and total organic-C, and information on natural and anthropogenic disturbances to sites and soils is not generally available.

This review therefore adopts a simplistic approach of assuming that SOM accumulations in natural tropical systems approximate to steady-state conditions and seeks to identify determinants of pool size. The consequences of temporal changes in inputs and rate determinants are then considered. Our objective is to identify the circumstances under which SOM in tropical soils at the ecosystems level is not primarily driven by physical factors and where biological processes involved in the turnover of litter and humus emerge as controls over SOM dynamics.

Climatic Factors

The standing crop of organic matter in tropical soils is determined by the net balance between the rates of litter inputs and decomposition, transfers to SOM pools, and the slower mineralization rates of this fraction. Little information other than aboveground production and total SOM accumulation is available for most ecosystems. Using these data, global climate patterns can be related to both plant production and decomposition, so that Post et al. (1982) were able

Table 1. Scales of resolution of litter decomposition and SOM accumulation (after Swift et al., 1979).

Scale of resolution	Environmental controls	Resource type and quality	Organisms	SOM pool
Ecosystem	Macroclimate, edaphic conditions, and soil texture	Total leaf, root, and wood litter	Total biota (95% microbial metabolism)	Total SOM
Population	Microclimate, soil structure, and gradients	Composition of litter by plant species	Functional groups and key animal species, total microbiota	Patch variation in total SOM
Organism	Microclimate, soil structure, and soil minerals	Cellulose, lignin, microbial cells, and products as energy sources	Microbial and animal species	SOM fractions
Molecular	Oxygen, water, and soil minerals	Specific substrate energy sources	Free and bound enzymes	Specific compounds

to superimpose contours of soil organic C pools on life zones defined by potential evapotranspiration and mean annual precipitation. The main trends in these relationships show that SOM accumulation is inversely related to latitudinal gradients of primary production and that soil organic matter accumulation in the tropics increases with altitude. A marked inflection of the SOM contours was found between the warm temperate and subtropical zones as a consequence of the dry season in subtropical forests having greater effects on litter production than on decomposition, resulting in low SOM accumulation. The results of this synthesis suggest that decomposition rates change faster as a function of temperature than does litter production and supports the thesis of Jordan (1985) that the imbalance between production and decomposition results in relatively lower SOM accumulations in tropical soils than in soils in higher latitudes.

In contrast to this emphasis on apparently uniform geographical patterns, Sanchez et al. (1982) and J. M. Anderson and Swift (1983) showed that there are no significant differences between the range of SOM contents found in temperate and tropical soils. Brown and Lugo (1982) adopted a similar approach to Post et al. (1982) in analyzing relationships between temperature and precipitation (T/P x 10^{-2}) with C pools in tropical forest biomass and SOM. Soil C showed a significant ($P = 0.05$) negative exponential relationship with T/P, but only 23% of the variance was explained by the regression. Much of the unexplained variance was contributed by sites with a T/P value of less than 1 (e.g., from 25°C/2500 mm and cooler wet forests), which had higher SOM contents than predicted. The forest life forms in this group include tropical and subtropical rainforest and montane forest representing approximately 25% of the area of the tropical forest biome (Brown and Lugo, 1982).

It would be incorrect, however, to necessarily assume from this study that SOM is less variable in dry tropical regions and therefore more predictable from climatic data. Log transformation of organic-C accumulations was not carried out and consequently the regression analysis was biased by data from the humid tropics (J. A. Jones, personal communication). As will be shown later in this chapter, significant variation in SOM accumulation can be imposed by site variables such as termite activities and soil texture. Furthermore, alternating phases of SOM depolymerization during wet seasons and polymerization at the beginning of the dry season can form more stabilized fractions under seasonal climates than under moister regimes (Duchaufour, 1977).

Site characteristics for several rainforest sites, shown in Table 2, illustrate the point that SOM contents can be highly variable within and between forests subject to closely similar climate conditions. The four forest sites in Sarawak, for example, have constant temperature and high rainfall regimes, but surface (0-100 mm) SOM concentrations range fourfold between sites and approximately twofold between 10 x 10 m plots representing extremes of soil conditions within the 1-ha alluvial forest and heath forest sites. Wide variation

Table 2. Characteristics of tropical rainforest sites where selected decomposition studies have been carried out.

Site and location	Altitude (m)	Annual precipitation (mm)	Dry season (months < 100 mm)	Monthly temp. range (°C)	Soil type	pH	Org-C (%)#
1. Sarawak (Gunung Mulu)							
Alluvial forest	50	5090	<1	26-28	org.	3.8	11
					min.	4.9	6
Dipterocarp forest	200-250	5110	<1	26-28		4.1	11
Limestone forest	300	5700	<1	26-28	min.	3.6	42
Heath forest	170	5700	<1	26-28	org.	3.7	19
						3.5	43
2. Sri Lanka (Sinharaja)							
Dipterocarp forest	518-548	3810-5080	<1	23-26	org.	5.0	5
3. Cameroon (Korup)							
Lowland rainforest	50	5460	3	24-30	sub.1*	4.5	3
					sub.2	4.7	4
4. Amazonia (San Carlos)							
Terra Firme	122	3565	<1	24-28		3.8	6
Caatinga	117	3565	<1	24-28		3.8	8
Bana	118	3565	<1	24-28		3.8	8
5. Jamaica (Blue Mountains)							
Ridge forest	1615	2500	1-2	9-24	mor*	3.0	47
Ridge forest	1600	2500	1-2	8-23	mull*	3.6	29
Wet slope forest	1575	2500	1-2	10-22		4.1	8
Gap forest	1575	2500	1-2	10-22		4.3	13
6. New Guinea (Marafunga)							
Ridge forest	2480	4000		8-18		5.9	23
Valley forest	4440	4000		8-18		5.9	16

* 0-100 mm, refer to text discussion for subplot or site descriptions.
# 0-100 mm, calculated as ignition loss x 0.6 where organic-C is not determined.
1. Proctor et al. (1983), Anderson et al. (1983), Newbery and Proctor (1984).
2. Mahareswaran and Gunatilleke (1988).
3. Gartlan et al. (1986).
4. Cuevas (1983).
5. Tanner (1981).
6. Edwards and Grubb (1982).

in SOM concentrations is also shown for groups of sites in Jamaica and New Guinea (see Table 2). Brown and Lugo (1982) concluded that the sites with higher SOM contents than those predicted by the climate model represent strongly modifying effects of soil conditions on C mineralization rates (or the stability of the humus products). At the other extreme of environmental conditions and SOM accumulation, M. J. Jones (1973) noted that the mean organic-C concentration of 605 well-drained savannas in West Africa was 0.68%. In multiple regressions, soil clay content accounted for 47.5% of the variance and rainfall 57.2%, and regression coefficients were interpreted to indicate slower decomposition rates of SOM associated with clays than with sandy soils.

Edaphic Effects

The influence of soil physicochemical conditions on SOM dynamics will not be considered in detail here except to provide a context for considering organism activities.

Organic-C concentrations in tropical soils generally exhibit a rapid decrease below 5-10 cm and then a steady decline to less than 1% at a depth of 50-100 cm (Sanchez, 1976). Root biomass and production generally parallel this spatial pattern in the soil profile. In Andosols, however, SOM typically does not show a rapid decrease with increasing depth in the profile (Sanchez, 1976). Amorphous alumina in Andosols is known to complex and stabilize organic matter, but the high SOM concentrations at depth relative to organic matter inputs implies that there is also redox gradient interacting with SOM stabilization by the clay.

Physical protection of organic matter from microbial attack and/or complexing also occurs in soil aggregates (Oades and Ladd, 1977; D. W. Anderson, 1979; Paul, 1984). Tillage practices, which increase the aeration of soils and break down macroaggregate structure, result in increased oxidation of SOM in temperate and tropical soils (Douglas and Goss, 1982; Alegre and Cassel, 1986). Tillage and cropping systems also alter the biological processes involved in aggregate formation, such as particle adhesion by microbial and earthworm mucopolysaccharides, and the binding effects of roots and hyphae. Thus the location of fine-root inputs within clay soils and the formation of soil aggregates by fauna may be important processes contributing to SOM dynamics in some sites. Changes in land use, such as conversion of forest to grassland or arable cultivation with different cropping systems, will alter the balance between organic-C stabilization and mineralization as these biological processes are disrupted.

Conditions of restricted aeration under waterlogged conditions inhibit C mineralization and are generally recognized to promote the development of organic soils in tropical lowlands where impeded drainage has arrested processes of decomposition and humification (Mohr and van Baren, 1954). The depolymerization of lignin requires a predominantly aerobic environment

Dynamics of SOM in Tropical Ecosystems

(Zeikus, 1981), but hydrolysis of cellulose and oligosaccharides can proceed under anaerobic conditions. Figure 1 shows that waterlogging has a transient, bimodal effect on C mineralization. Once anaerobic conditions are established, C mineralization increases but the effect is reduced with time, and depth in the profile, reflecting the reduced availability of C compounds in litter and SOM.

The deep accumulation of fibrous organic matter in Southeast Asian heath forest in Sarawak (Table 2) is promoted by waterlogging, but the high

Figure 1. Carbon mineralization from soil sampled at three depths in an Andosol and incubated at 28°C for 35 days different gravimetric moisture contents. Traces of carbon mineralization at day 2 and days 32-35 reflect the reduction of microbial-available carbon with time and depth. The bimodality of the curves shows that waterlogging initially inhibited carbon mineralization but that rates recovered when anaerobic processes were established. Similar bimodal patterns of carbon mineralization in relation to waterlogging were recorded for soil samples from Vertisols. (Adapted from Legay and Schaefer, 1981.)

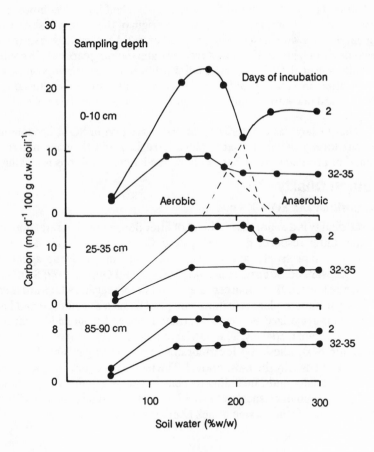

surface SOM accumulation in dipterocarp forest on well-drained slopes may be a function of litter quality with exceptionally low phosphorus content (Proctor et al., 1983). In the forest on limestone, however, the inhibition of humification processes by active calcium carbonate may account for surface accumulations of organic matter as occurs in rendzinas (Duchaufour, 1977). On the other hand, slightly elevated areas of the alluvial forest had acid, organic soils in comparison with surrounding gleyed alluvial soils of lower organic content. Newberry and Proctor (1984) showed that exchangeable Ca is more than 12 times higher in the gley (6.2 meq 100 g^{-1}) than in the podzolic areas (0.5 meq 100 g^{-1}) on the alluvial forest site and suggested that the more acid soils are just above the maximum level of flood water inputs from nearby catchments on limestone. It is evident that different, unquantified edaphic effects are important for determining the SOM dynamics of these sites.

Site-specific effects are also evident where comparative studies are made with the same resource type or standard. Tanner (1981), for example, demonstrated significant site differences between leaf litter decomposition in two montane forests irrespective of the site origin of the material; slower rates of decomposition were associated with higher SOM accumulation at the mor site with lower pH, higher CEC, and greater water-holding capacity. In contrast, J. M. Anderson et al. (1983) found small site differences in decomposition rates of mixed litters in Sarawak, although *Parashorea* litter from the alluvial forest showed highest decomposition rates in the litter layers of the waterlogged heath forest soil.

These examples emphasize that wide variation in SOM accumulation can occur locally under the same climatic regime and they are not readily explained or predicted when information on edaphic conditions is lacking.

Resource Quality

Decomposition and SOM Formation

Resource quality is a major determinant of litter decomposition rates and SOM formation. Litter resource quality may be defined in terms of C and nutrient availability to saprotrophs and the concentrations of modifying agents that inhibit organism and/or enzyme activities (Singh and Gupta, 1977; Swift et al., 1979). In plant litters, the C sources range from simple sugars, which are utilized by most organisms, to plant structural compounds such as cellulose and lignin, which are depolymerized by a more restricted range of fungi and bacteria and by still fewer animal species. The nutrients theoretically include all elements required for biosynthesis by invertebrates and microorganisms, but N and phosphorus are usually the most critical. The modifiers comprise a wide range of secondary compounds, including terpene, alkaloid, tannin, and polyphenol species, and xenobiotics such as pesticides. Lignin also acts as a modifier by masking cellulose from enzyme attack (Zeikus, 1981).

The same concept of resource quality can be applied to the groups of SOM constituents that are more or less resistant to enzyme attack, with edaphic conditions acting as modifiers.

When fresh litter enters the soil, initial rates of C mineralization are usually high, under favorable climatic conditions, and then decline as labile C compounds are exploited by the biota. Both synthesis of new compounds by microorganisms and complexing of metabolites are concurrent with mineralization, so that, with time, the more recalcitrant plant structural materials and microbial products comprise a greater proportion of the residual mass. Cheshire et al. (1974), for example, showed that after incubation in soil for 443 days only 2.3% of the original xylose in hemicellulose remained but mannose, which is a major constituent of fungal cell walls, increased by 450%. The walls of pigmented fungi are highly resistant to decomposition and are an important contributor to SOM formation (Wagner and Mutakar, 1968).

Studies using ^{14}C-labeled plant and microbial tissues have shown that C is incorporated into the fulvic acid, humic acid, and humin fractions of SOM within a few weeks of incubation in soil (Paul and van Veen, 1979). Some component of the plant C may also be combined in these SOM fractions after only partial modification by microorganisms. J. P. Martin and Haider (1977) found that the bulk of ^{14}C-labeled ring structures from partially degraded lignin precursors could be recovered in humic acid.

Carbon-dating methods have shown that humic acids have mean residence times of thousands of years compared with hundreds of years for the humin fraction. The fulvic acid fraction contains polysaccharides with fast turnover times and are generally contemporary with the humin fraction or are of more recent origin (Paul and van Veen, 1979). If fulvic acids move down the soil profile, however, they can be among the oldest fractions (Goh et al., 1976).

SOM accumulation is thus not simply an inverse function of litter decomposition rate, although the two processes are interlinked through resource quality effects. Edaphic conditions can act differentially on both processes. Hence waterlogged conditions inhibiting decomposition inhibit SOM formation and cellular "raw humus" material may accumulate. Conversely, high stabilization of C from rapidly decomposing resources can occur in clays.

Resource Quality Determinants of Litter Decomposition Rates

Many studies have identified lignin, N, and total phenolic concentrations as key resource quality determinants of litter decomposition rates but the literature is conflicting in terms of the relative importance of these parameters as regulatory variables.

The lignin concentration and the lignin/nitrogen (L/N) ratio of plant materials (Melillo et al., 1982; Berendse et al., 1987) generally correlate with the decomposition rates of major resource types, such as wood and leaves, and differences within resource types (Berendse et al., 1987; Cuevas and Medina, 1988). La Caro and Rudd (1985), however, found lignin concentrations alone the best correlate with leaf litter decomposition in El Verde, Puerto Rico, and nutrient concentrations appeared to have no influence on decomposition rates. On the other hand, Schaefer et al. (1985) found that mass losses of different litters in a New Mexico desert were not explained by lignin or nutrient concentrations.

Evidence for the modifying effects of secondary plant compounds, notably polyphenols, on invertebrate and microbial activities is also conflicting (J. S. Martin and Martin, 1982; Zucker, 1983). Toutain (1987) showed that moribund leaves that contain tannin have up to 70% of cytoplasmic proteins complexed with tannin and with potentially reduced availability to decomposers. This hypothesis was first proposed by Handley (1954) to account for N immobilization in mor forest soils in temperate regions but is inconsistent with the results of protease assays carried out by van der Linden (1971), which showed proteolysis declined with time in decomposing litter. On the other hand, decomposition rates of two legume mulches, *Inga edulis* and *Erythrina* sp., at Yurimaguas, Peru, were reported as more closely related to polyphenol contents than to L/N ratio (C. A. Palm, personal communication). But a fivefold difference in polyphenol concentrations was not reflected by decomposition rates of *Ficus* sp. and *Parashorea* sp. in the Gunung Mulu alluvial forest; the L/N ratios for the two litters were more closely related to mass losses from litter bags (Table 3). Three reasons can be put forward to account for the inconsistency of these modifier effects.

First, it would appear that C, nutrient, and modifier parameters defining resource quality may operate as another series of hierarchical controls, or limiting factors, according to the relative importance of resource, organism, and physical controls on decomposition. Thus the marked effects of total phenolics on legume-mulch decomposition rates may reflect a situation in which low lignin and high N concentrations have released decomposition rates from these constraints so that the phenolics can act as rate determinants.

Second, there are qualitative differences in the biostatic effects of different phenolic compounds (Benoit and Starkey, 1968; Levins, 1971; J. S. Martin and Martin, 1982; Zucker, 1983), which have not generally been investigated by soil ecologists.

Third, hydrolyzable phenols leach rapidly from leaf litter so that site differences in the precipitation regimes may have separated the modifiers from the litter-decomposing organisms. The modifying effects of the phenolics may therefore be expressed in relation to SOM formation and decomposition at sites divorced from the parent litter. In tropical Andosols many plant species have root and leaf litters rich in tannins (Darici et al., 1988). Some of the stabilized organic matter at depth may result from the effects of soluble phenolics leaching down the profile, which directly or indirectly affects microbial processes. Also,

Table 3. Leaf-litter resource quality, disappearance rate, and SOM accumulation in four rainforests in Gunung Mulu National Park, Sarawak (see Table 2). Litter decomposition was estimated from mass losses of leaves in coarse- (20-40 mm) and fine- (40 μm) mesh litterbags and the turnover coefficient (k_L) was calculated from litterfall and standing crop values. (Data from J. M. Anderson et al., 1983.)

Litter type and site	Mineral elements (% d.w.) Ca	P	N	Lignin (%)	L/N ratio	Polyphenols (% tannin equiv.)	Litterbag mass loss (% yr) fine	coarse	Turnover coeff. (k_L) leaves	total	SOM[a] (Mg ha⁻¹)
Alluvial forest mixed litter	2.0	0.03	0.8	31	39	1.7	53	60	1.8	1.7	230
Ficus sp.	5.7	0.07	1.0	27	27	0.5	74	97	–	–	–
Parashorea sp.	1.7	0.03	0.8	28	35	2.5	63	98	–	–	–
Dipterocarp forest mixed litter	0.4	0.01	0.8	38	39	1.9	49	56	1.7	1.3	198
Heath forest mixed litter	0.5	0.01	0.4	40	100	2.3	59	61	1.4	1.2	318
Forest on limestone mixed litter	3.5	0.04	1.1	39	35	1.5	65	52	1.7	1.5	164

a. Sampled at the 0-30 cm depth in alluvial, dipterocarp, and heath forests, and at the 0-11 cm depth in forest on limestone.

allophane soils bind enzyme/protein complexes and protect them from biodegradation (Darici et al., 1988).

Inter- and intra-specific variation in the quality of litter resource types (leaves, woody materials, roots, etc.) is a function of nutrient availability to plants. If nutrient availability is high, fast-growing plants produce nutrient-rich litters that decompose rapidly and nutrient cycling is largely outside the plant through the litter/SOM pathway. At the other extreme, Vitousek (1984) has shown that in nutrient-stressed forests the production of leaves or wood is more efficient per unit of limiting nutrient than it is on high fertility soils. This is achieved by the selection of plant species that have low nutrient requirements; for example, the N requirement of temperate deciduous trees is approximately twice that of conifers in the same region (Cole and Rapp, 1981). In addition, nutrients may be translocated more efficiently from senescent tissues before abscission under nutrient-limited conditions (Charley and Richards, 1983; Vitousek, 1984). Chapin and Kedrowski (1983) contend, however, that there is no evidence for this phenomenon and that low nutrient concentrations in plant litter on nutrient-limited sites is a consequence of structural compounds diluting the minimum nutrient capital available to the plant. Higher levels of fine-root production (Gower, 1987) and evergreen, scleromorphic leaves (and hard woody tissues) are associated with drought and/or nutrient-stressed habitats (Werger and Ellenbroek, 1978; Sobrado and Martin, 1980). There is also an extensive literature suggesting that plants in nutrient-stressed habitats are likely to have higher concentrations of carbon-based secondary compounds against insect attack than plants on nutrient-rich sites, although the causal basis of this relationship is unproven (Chapin, 1980).

Irrespective of the mechanisms, it could be expected that plant materials entering the decomposer system under nutrient-stressed conditions may have low resource quality characteristics and decompose slowly and have a high potential for the development of stabilized SOM pools. This may be a contributary factor in the development of Spodosols under tropical forest, but there is little evidence for this relationship at the ecosystem level for natural plant communities on Oxisols and Ultisols of low inherent fertility. Within these systems, however, there is considerable variation in litter resource quality and decomposition rates reflecting different nutrient demands and recycling strategies of different plant species. For example, the low-quality, mixed alluvial forest litter in Sarawak decomposed more slowly than did the high-quality *Ficus* and *Parashorea* leaves from the same site (see Table 3). Conversely, areas under canopy dominance by *Mesua nagasari* and *Dipterocarpus cylindricus* on comparatively fertile soils in the Sinharaja MAB Forest Reserve, Sri Lanka, contained twice (8%) the average organic-C concentrations of soil (0-10 cm) under the surrounding forest (J. M. Anderson and I. U. A. N. Gunatilleke, personal communication). Incipient podzolization under the canopies of individual tree species was also recorded in temperate regions (Crampton, 1982). The determinants of this patchiness are unknown but would appear to be a function of the ecological characteristics of the tree species rather than climate, edaphic, or specific organism activities.

At present we have little quantitative understanding of how resource quality parameters interact, and conventional wisdom is strongly biased toward temperate studies in which lignin, N, and polyphenol concentrations are often covariates. In tropical vegetation there is greater independent variation in these parameters, as illustrated in Table 3, where litters with high lignin vary in total-N, total polyphenol, and base cation concentrations, with unknown interactive effects on decomposition processes and SOM formation.

Organism Activities

Situations exist in which particular groups of soil organisms exert a strong enough influence to overcome the effects of climatic, edaphic, and resource quality factors and thereby alter organic matter dynamics. The major organism groups with demonstrable effects on SOM include plant roots and soil invertebrates, which interact with and influence microbial populations.

Roots

Roots have several complementary physical effects on soil components. Roots, and associated mycorrhizal fungi, play important roles in binding soils and soil aggregates, thus impeding soil erosion. Soil aggregates and the underlying parent mineral strata are also disrupted by root penetration. Root channels provide avenues for movement of soil particles, water and solutes, and various members of the soil fauna. Deep penetration of roots, e.g., those of grasses and many tree species, also serves to transport organic C deep into the soil. These processes are considered further in Sanchez et al. (Chapter 5, this publication).

Root impacts on soil processes relate to several source/sink phenomena. Roots may be a C sink when deposited in deeper soil layers, but they also serve as an energy source for microbial and invertebrate populations. Soil fauna, such as termites and earthworms, ingest and process the microbial and root C and redistribute it through the soil matrix.

Fine roots can act as regulators of SOM dynamics as well as sources of organic matter (Jager, 1971). Root adherence to litter was found by Cuevas and Medina (1988) to accelerate decomposition and nutrient release in a Tierra Firme forest in the Amazon Basin. Billes and Bottner (1981) also demonstrated that C mineralization and humification rates of different SOM components are related to the phenology of wheat plants. Using ^{14}C-labeled root residues they demonstrated a two-phase decomposition process. There was an initial rapid flux of CO_2 output in both systems, then a decreased CO_2 flux, which coincided with increased root growth that continued to the flowering phase. After flowering, rates of CO_2 output increased until plant maturation. Thus in the rhizosphere, exudates and debris from growing roots intensify the biodegradation of organic matter and synthesis of microbial products (Helal and Sauerbeck, 1986). Labile C inputs to the soil from exudates and exfoliates can account for 18-25% of C fixed by the plant (Barber and Martin, 1976). Additional labile C, for instance extramatrical mycorrhizal C, may account for up to an additional 6-10% of root-associated C (Kucey and Paul, 1982). Symbiotic N fixers (e.g., rhizobia and actinorrhiza) can contribute a further 5-10% of the root-associated C to the soil matrix (Pate et al., 1979; Pang and Paul, 1980). After flowering in annual plants, with the onset of senescence, root inputs are reduced and mineralization of accumulated microbial products intensifies with increased decomposer activity.

Soil Invertebrates

The presence and population size of key animal groups or species can significantly affect the pathways and processes of C mineralization in ecosystems. The effects of grazing by vertebrate herbivores on the quality and quantity of C inputs to grassland soils is well documented, and the removal of crop residues or litter (as mulch or fuel) from natural systems will have predictable effects on soil C balances. But the effects of invertebrates on soil properties and processes are poorly quantified, although the biomass per unit area of soil invertebrates, notably earthworms and termites, often exceeds that of vertebrate herbivores in tropical ecosystems (Lee and Wood, 1971).

Saprotrophic soil invertebrates contribute directly and indirectly to soil processes. The direct effects involve C and nutrient mineralization through metabolic processes and elements recycled through biomass turnover. Numerous studies in temperate and tropical regions have shown that the soil fauna biota directly metabolizes no more than 10-20% of the total C flux through soils (Phillipson, 1973; Luxton, 1982); the flux through termite populations being largely mediated by gut protozoa of bacteria or by the fungus gardens of Macrotermitinae.

The indirect effects are the consequence of feeding and burrowing activities on the soil physicochemical environment for microbial processes and of root growth and nutrient uptake. The relative contribution of direct and indirect effects to soil processes is a function of body size, as this determines whether the activities of the animals are constrained by or can modify the structure of soil and litter habitats (J. M. Anderson, 1988a). Soil macrofauna, notably termites and earthworms, can therefore have disproportionately larger effects on energy fluxes, including soil water balances, than their direct contribution to community metabolism. These direct and indirect effects of termites and earthworms on SOM dynamics can be defined in terms of epigeic species, which are involved in processing litter on and in the top few centimeters of soil; endogeic species, which live in mineral soil and may be geophagous; and anecic species, which transfer soil and plant materials between habitats (Figure 2). The activities of these groups potentially affect the quantity, quality, location, and environment of microbial decomposition of litter and SOM formation.

Figure 2. Ecological groups of soil and litter invertebrates that affect organic matter distribution . (After Bouché, 1977.) The epigeic species process litter in the surface layers of the soil and passively effect the transport of particulate materials and solutes. Similarly, the endogeic species living in mineral soil horizons do not actively redistribute organic matter between soil and litter layers but may influence water flux pathways and leaching. The anecic species, however, may incorporate litter into the soil or may remove it from the soil system altogether.

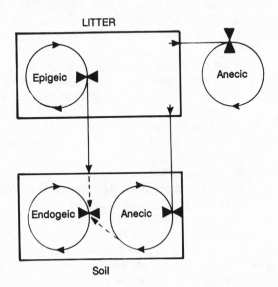

Details of termite and earthworm effects on soils have been reviewed elsewhere (Lavelle, 1988; Wood, 1988) and the following examples illustrate their potential significance on SOM dynamics.

Termites

There is considerable evidence that termites can have a major impact on litter decomposition and SOM accumulation and, thereby, on soil physical properties such as aggregation, structure, and permeability in tropical soils. Termites can be grouped according to whether they are epigeal (aboveground, usually tree dwelling) or hypogeal (belowground); whether they forage for wood, litter, or humus; and whether they construct distinct nests. The different groups have potentially different effects on SOM according to food type, population densities, and the distance over which organic materials are transported to the nests.

The activities of the fungus-cultivating Macrotermitinae have the most clearly documented effects on C and nutrient cycling in seasonal forests and grasslands of tropical Africa and Asia. Litter is transported to fungus gardens over distances from a few meters or less, in the case of *Microtermes* and *Ancistrotermes* spp., which have diffuse nest systems within the soil, to more than 50 m for mound-building *Macrotermes* spp.

In the seasonally wet tropics of Malaysia, termite populations at Pasoh forest were estimated at 3000-4000 m^{-2}, of which 60% were wood-feeders, 25% humus-feeders, and 15% litter-feeders (Matsumoto and Abe, 1979). Although termites consumed 16% of total aboveground litter production, up to 32% of daily litterfall was locally consumed by *Macrotermes carbonarius*, with the consequence that lower litter standing crops were recorded around *M. carbonarius* mounds than elsewhere on the forest floor (Matsumoto and Abe, 1979). This group of termites was scarce or absent in the high-rainfall Mulu forests, considered above, and termites did not contribute significantly to litter decomposition in these sites (Collins, 1983).

In the ungrazed Southern Guinea savanna in Nigeria, West Africa, Ohiagu and Wood (1979) estimated that termites consumed 980 kg ha^{-1} yr^{-1} out of an annual grass litter production of 1226 kg ha^{-1} and that 790 kg ha^{-1} were removed by termites during a 4-month period when rainfall was negligible. Estimates of woody-litter consumption by termites in other African savannas range from 60% in Nigeria, 86% in Ghana, and 91% in Kenya (Buxton, 1981); Boutton et al. (1983) reported 64-70% of grass litter and 30-36% of woody litter removed by a *Macrotermes* sp. in two Kenya grasslands. Gupta et al. (1981) showed in an Indian grassland that an *Odontotermes* sp. selects grasses and forbs, with up to 100% attack on some species during the dry season. Figures of around 17-18% woody litter consumption by other groups of termites were reported from South Australia and Arizona (Buxton, 1981).

Humus-feeding termites occur in semiarid grasslands but reach maximum population densities and species diversity in the wet tropical rainforests of Africa, Asia, and South America where litter-feeding termites are less abundant. Some species construct mounds on the soil surface, predominantly of fecal material, but the majority of species have diffuse nest systems in soil with galleries lined with fecal material. Humus-feeders exploit SOM fractions and do not attach intact litter. For example, J. M. Anderson and Wood (1984) showed that the organic matter and silt/clay content of the mounds of two species in Nigeria increased as a function of these fractions in the surrounding soil. The termites were clearly exploiting the SOM fractions intimately associated with mineral material rather than the fine-litter fractions. Little work has been done on the digestive physiology of humus-feeders, but Bignell et al. (1983) suggested that depolymerization of stabilized SOM compounds is carried out by alkaline hydrolysis at pH 9-11 followed by the assimilation of products from a bacterial "chemostat." As a possible consequence of these extreme gut conditions, the available-P content of the fecal mounds was up to 16 times higher than surface soil concentrations. Wood et al. (1983) calculated

Dynamics of SOM in Tropical Ecosystems

that humus-feeder populations of 2369 m^{-2} in a Nigerian forest could turn over 30% of the top 25 cm of soil per year.

It is evident that termites can have a major impact on litter decomposition and SOM dynamics. Pomeroy (1983) found that SOM accumulation was lower within the foraging range of large termite mounds in Kenya. At high mound densities, where most of the aboveground litter is removed by termites, J. A. Jones (1988) also found that soil organic-C concentrations were 10-100 times lower than where termites were scarce or absent. No studies have experimentally demonstrated the effects of Macrotermitinae on SOM accumulation, but the elimination of termites with pesticides in a New Mexico desert (Elkins et al., 1986) resulted in increased SOM accumulation and was associated with major changes in the hydrology and herbaceous vegetation of the area. The combined effects of litter- and humus-feeding termites could result in lower SOM contents of soils than might be predicted from the quality and quantity of aboveground litter inputs. Trapnell et al. (1976) investigated the effects of fire protection on a Zambian woodland soil over a period of 23 years. Organic-C contents (0-15 cm depth) of the unburned plots (0.85%) were similar to those of the burned plots (0.97%), and it was suggested that the increase in litter- and humus-feeding termites in the fire-protected plot had complementary effects to fire in removing aboveground litter inputs to the soil. Under these extreme conditions, SOM accumulation may largely depend upon the quality, quantity, and location of root inputs.

Earthworms

The dramatic effects of temperate earthworms on litter decomposition, SOM formation, and the redistribution of organic matter in soils (Lee, 1985) are particularly associated with the activities of anecic species. This ecological group is less important in many tropical soils where epigeic and endogeic humus-feeding species are more representative of earthworm communities (see Figure 2).

In West Africa, seasonal drought may limit earthworm distribution and they are not found where annual rainfall is less than 800 mm and the dry season extends to 4-5 months. Earthworms may also be limited in tropical forests with less than 1300-1600 mm rain and where evapotranspiration may dry out the soil within a short period after rainfall. Consequently, earthworms are usually abundant in grassland (savanna and pasture) systems with greater than 1200 mm annual rainfall and forests with greater than 1800 mm rainfall (Lavelle, 1988) (Figure 3). The Amazon rainforest, and South America in general, may have large populations of truly anecic species that make up 40% of earthworm biomass at San Carlos de Rio Negro, Venezuela, (Nemeth, 1981) and 45.6% in peach palm plantations on Ultisols at Yurimaguas, Peru, (P. Lavelle and Pathanati, personal communication). In contrast, West African and Central American rainforests are dominated by endogeic species feeding on SOM

Figure 3. Latitudinal gradients of resource use by earthworms. CF = coniferous forest; AG = alpine grasslands; DF = deciduous forest; MF = Mediterranean forest; P = temperate pastures; TRF = tropical rainforest; DTF = subhumid (deciduous) tropical forests; DS = dry savannas; MS = moist savannas. (After Lavelle, 1988.)

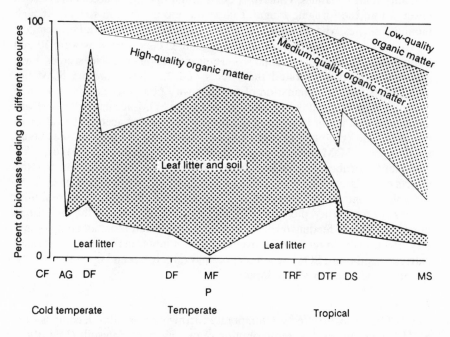

(Fragoso and Lavelle, 1987). Finally, small populations of earthworms have been reported from Southeast Asian forests (J. M. Anderson et al., 1983; Collins et al., 1984) with climatic and soil conditions similar to those in forests within other geographic regions where earthworms are more abundant. The reasons for this are still unclear.

The largest populations of earthworms, with biomass values of 300-600 kg fresh weight ha^{-1}, have been reported from moist grasslands in India, West Africa, and South America, and 1-2 Mg ha^{-1} in improved pasture in Peru colonized by the peregrine species *Pontoscolex corethrurus* (Lavelle, 1988) and up to 3 Mg ha^{-1} in pastures of Vertisols on the Island of Guadeloupe (Barois et al., 1988). In tropical grasslands, such as at Lamto, Cote d'Ivoire, geophagous earthworms may ingest daily 5-36 times their own weight of soil, depending on species, individual size, soil temperature and moisture conditions, and water-soluble organic matter contents of the soil (Lavelle, 1984). Consequently, up to 1200 Mg dry weight soil ha^{-1} yr^{-1} may be processed, containing 33% of the overall SOM content of the soil down to the 40-cm depth and 60% of SOM in the top 10 cm of the profile (Lavelle, 1978). One consequence of this turnover

Dynamics of SOM in Tropical Ecosystems

is the digestion of stabilized SOM by geophagous species achieved by priming the activities of gut bacteria with mucus-containing labile C compounds (Barois and Lavelle, 1986; Martin et al., 1987). Microbial activity may be still enhanced for 4-8 days after voiding, but C-mineralization in casts soon declines and after 420 days of incubation (at 27°C and pF 2.5) total C mineralization in casts was 3.3 times greater than in a control soil (Martin, personal communication). In contrast, Shaw and Pawluk (1986) showed in laboratory experiments that the intimate mixing of organic matter and clay minerals by a temperate endogeic earthworm increased the stabilization of organic-C in soil aggregates, but this has not been demonstrated to contribute to SOM accumulation under field conditions.

Changes in Litter Inputs and Controls over SOM Dynamics

The previous discussion assumes steady-state conditions for litter decomposition rates, stabilization of organic C in SOM pools, and mineralization rates of SOM. Each step in these transformations of organic C is regulated by a suite of biotic (O), physical (P), and chemical (Q) parameters, which in reality, vary within and between plant/soil systems in their relative importance and effects. The observed levels of SOM accumulation are the net effects of changes in these parameters over different spatial and temporal scales.

Model of Organic Matter Dynamics in the Plant/Soil System

The integration and interaction of the OPQ parameters that regulate organic matter dynamics at all levels in the plant/soil system are illustrated in Figure 4.

In the plant system of this model, nutrient availability and climate (and fire) act on the plants to determine the quality of litter and the partitioning of inputs to the above- and belowground litter compartments of the soil system. The organism parameters regulating these inputs include herbivores, which can affect both the quality and location of litter above- and belowground. Transfers of organic matter from aboveground (L1) to belowground (L2) litter and SOM pools are determined by the rate at which litter decomposes on the soil surface, the export of litter by animals, and the transport of soluble and particulate materials into the soil compartments. Animals (including man) may actively transfer organic matter from aboveground to belowground litter pools or may indirectly facilitate physical transport processes (water and gravity) by litter comminution or by changing soil structure and water balances (J. M. Anderson, 1988b). As C losses increase from L1 through mineralization and export, the inputs and stabilization of organic matter in L2 are of increasing importance for maintaining optimum SOM pools. Accelerating or diminishing L2 and SOM turnover rates can alter nutrient availability to plants and hence feedback to litter inputs and resource quality.

The O, P, and Q parameters of this model can be defined at different spatial scales from local patches in the vegetation mosaic dominated by a plant species, termites, or local edaphic conditions to mosaics of tropical vegetation and soil types within the landscape. The parameters will also change with time from seasonal events, through intermediate time scales of decades as plants establish and die, to the longer time scales of ecosystem succession, changes in land use, and variation in climate.

Figure 4. Schematic representation of organism (O), physical environment (P), and resource quality (Q) controls over organic matter transfers in the plant/soil system. All transfers are regulated by a combination of OPQ variables: in the plant system these variables are plant species (O), temperature and moisture conditions (P), and nutrient availability (Q). The combination of these variables determines the quality and quantity of above- and belowground inputs to the soil system. The organisms processing litter and SOM include invertebrates and microorganisms but man can also affect inputs and exports from the system. Transfers of organic matter between the above- and belowground compartments may be effected actively (by invertebrates and man) or passively (by microorganisms and invertebrates), as illustrated in Figure 2. These transfers are also affected by climatic conditions and resource quality. The organisms regulating carbon mineralization include plant roots. The size of litter and SOM pools is therefore determined by the balance between carbon inputs and carbon losses by export, transfers, and mineralization.

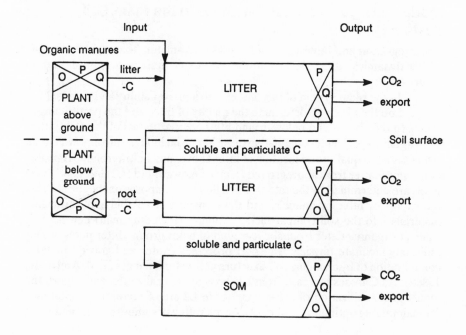

Dynamics of SOM in Tropical Ecosystems

The spatial and temporal changes in plant communities will leave a legacy of higher or lower rates of stabilization or mineralization of SOM fractions according to the resource quality of inputs (particularly lignin), changes in termite, earthworm, or herbivore populations, land use by man, and so forth. The frequency, area, and duration of events will determine whether they contribute to the stabilized SOM pool for the system or only have transient effects on more labile SOM pools.

Spatial and Temporal Dynamics in Natural Systems

Tropical forests comprise mosaics of disturbed patches formed by tree-fall events and regeneration. Inputs of litter to patches will therefore vary by orders of magnitude from occasional, massive inputs of low-quality, woody litter associated with gap-formation and to shorter term changes in the quality and quantity of leaf-litter inputs according to light intensity, nutrient availability, and species composition of regrowth vegetation. Variation in SOM accumulation under different tree species on the same parent soil in natural forests, or under monocultures in forestry plantations, as considered above, illustrates species effects on C fluxes that are of unknown duration and contribution to stabilized SOM pools. Patchiness of invertebrate effects in litter selection by termites and wide extremes of earthworm activities under different tree species (Gould et al., 1987) may also contribute to this dynamic mosaic of balances between litter inputs and SOM accumulation. It has long been recognized that an age structure for gaps exists, but the dating of gaps has been limited to known tree-fall events. Martinez-Ramos et al. (1988) have used palm regrowth to date gaps in a Mexican rainforest and have provided a "life table" for gaps up to more than 80 years old with a mean rotation period of 53 years. The impact of these dynamic processes on transfers of organic-C to SOM pools and the relative sizes of the active, slow, and stabilized pools will affect sampling procedures, unexplained residual variation, and interpretation of the response of the system SOM pools to larger perturbations such as clearfelling and subsequent changes in SOM.

Effects of Changes in Land Use

Clearance of natural vegetation for agriculture changes the amounts and qualities of organic matter inputs to soils, soil temperature and moisture regimes, and biological processes affecting litter decomposition and SOM. Tillage practices also increase organic matter oxidation by disrupting soil aggregates, exposing new surfaces to microbial attack, and changing the redox conditions at depth in the profile. Soil fauna populations are highly susceptible to these environmental changes and the numbers of invertebrate species, total population densities, and functional groups (notably the epigeics) decline rapidly when natural vegetation is cleared and intensively cultivated (Lavelle, 1987). Conversion to pasture, traditional cultivation, and other conservation tillage practices have selective effects on earthworm and termite populations adapted to the new conditions. Studies in temperate grassland and minimum-tillage systems, where earthworms have been experimentally introduced or eliminated, have shown significant effects of earthworm population densities on the distribution and turnover of SOM (Lee, 1985). Comparable, manipulative studies have not been carried out widely in tropical systems to quantify these roles of soil fauna. But there is considerable evidence of termites and earthworms affecting water infiltration rates, soil porosity, nutrient cycling, and root distribution, which are likely to indirectly affect short-term SOM dynamics via plant growth and microbial activities (Lavelle, 1987; J. M. Anderson, 1988a). Perfect et al. (1980), for example, found that spraying cowpea with DDT over a 4-year period doubled grain yield in a Nigerian bush-fallow rotation with no fertilizer inputs. But the decline in yield over the cropping period was faster in the treated plots and was particularly marked in a subsequent maize crop. The effect was not attributable to greater nutrient depletion in the more heavily cropped plots. It was concluded that differences in soil fertility arose through supression of earthworm effects on soil properties which had little effect on population densities. The cumulative long-term effects of changes in soil fauna populations, pathogens, and agrochemicals on SOM in tropical agricultural systems are unknown at the present time.

The direction of change in organic matter in soils brought under cultivation depends upon the previous organic matter level as well as the cropping system (Juma and McGill, 1986). In general, the organic matter contents of tropical soils under continuous cultivation fall to 30-60% of the corresponding values under natural vegetation within a few years (Young, 1976) but the rate of decline can be reduced or arrested by different management practices. Ayanaba et al. (1976) showed that organic-C and N declined rapidly 2 years after clearing forest on Andosols in Nigeria, but the decline in C was slower, and in some cases organic matter increased, where low-quality maize residue was retained on site. Grasses and legumes in rotation decrease losses of SOM through high belowground primary production and continuous pasture increases SOM content, possibly because decomposition under unfertilized grass may be limited by N availability (Juma and McGill, 1986). In these sites the slow and passive SOM pools may be small and organic-C is maintained by a rapid turnover of a large active pool.

In contast, Buschbacher et al. (1988) found no differences in SOM between 8-year-old abandoned pastures in Brazil, which had previously been subjected to extremes of management intensity involving burning, stocking, and woody-residue removal. In this case the passive forest SOM pool appears large and may be masking the C dynamics of the smaller active fraction. This is illustrated by Cerri et al., (1985) who used $\Delta\ ^{13}C$ to investigate the turnover of soil organic matter after conversion of forest dominated by C_3 species to C_4 sugarcane: half of the initial soil C remained 50 years after conversion.

Dynamics of SOM in Tropical Ecosystems

Allen (1985) showed major differences in organic-C losses (and associated total-N and CEC) from a range of soil orders after forest clearing. Absolute losses were greater in soils on young, less-weathered soil parent materials, but relative losses were significantly greater in soils on more highly weathered soil parent materials. This suggests that even in the same climate, SOM in Andosols, Mollisols, and some Inceptisols may be more resilient than in Oxisols or Ultisols under cropping or forest plantations in the tropics.

Variation in Controls over SOM Dynamics in Time and Space

The period and intensity of these changes represents a time dimension to the hierarchy of controls over decomposition and SOM turnover discussed in the previous section. At a geographic scale, gross changes in SOM when areas of a landscape are cleared for a permanent change in land use can be related to vegetation type, climate, and soil type.

Alternation of arable cropping systems and extended grass or forest fallows involve ecosystem-level changes in the quality and quantity of resource inputs and conditions for SOM formation and mineralization. Shifts in the importance of invertebrates in these processes occur between the cropping and fallow phases.

Different agricultural practices emphasize different biological controls on SOM dynamics. In minimum-tillage systems resource quality and fauna effects can be important factors contributing to soil fertility (Swift and Sanchez, 1984; Lal, 1987), whereas resource quality effects on microbial processes are the main biological controls over SOM dynamics in intensively tilled arable systems.

Soil organic matter dynamics are fairly predictable at the extremes of landscape and intensively cultivated systems using current simulation models. But in complex natural systems and agroecosystems involving spatial and temporal changes in plants, animals, and soil conditions, there is little understanding of the scales at which biological processes operate and how they integrate into the functioning of the whole system. An understanding of how to manage soil biological processes to promote optimum conditions of SOM formation and nutrient release to plants would provide important management options for tropical agriculture (Swift, 1986). Studies on tropical soil processes can complement temperate studies by providing different combinations of climatic, resource quality, and organism parameters toward this understanding of the importance of biological controls over SOM dynamics.

Conclusions and Recommendations

Much of our panel discussion focused on ways to strengthen the sections concerning the effects of organisms on SOM. There was also concern that interactions between organisms and management systems were addressed inadequately; but it was concluded that this represents a lack of appropriate information on the roles of organisms in tropical soil processes.

Theme Imperative: An appropriate overall imperative for this group is closely allied to that of Chapter 2; we need to understand the regulation of SOM pools and turnover by climate, soil, management and *biological* factors, and the implications for human use of tropical soils.

Research Imperatives:

1. There is evidence that populations of soil organisms can demonstrably control soil organic matter in some tropical environments. We need to characterize where (e.g., along which environmental gradients) such effects are important and to conduct appropriate experiments designed to isolate the effects of biological factors. It may be possible to make use of biogeographic patterns of the distribution of soil organisms in this analysis.

2. Carefully controlled experiments should be designed to evaluate the effects of management practices on populations of soil organisms and, more importantly, the consequences of changes in functional groups of organisms on SOM transformations in managed systems. Such experiments should involve additions and removal of groups of organisms in both natural and managed ecosystems.

3. Studies should be developed to evaluate food-web interactions in managed tropical systems. The importance of soil fauna and roots in regulating microbial populations, activities, and dispersal has been the subject of much discussion but with rather less quantification to date.

4. While there has been substantial research on the effects of organic matter inputs on rates of litter decomposition, there is much less information on the effects of organic matter quality on the stabilization of SOM. Initially, this work should be carried out in tree plantations (holding climate and parent soil conditions constant) so that the chemistry of organic matter inputs and cumulative effects on SOM accumulation can be determined.

5. The effects of roots on the turnover of SOM pools would reward substantial effort. The placement and timing of root inputs is important; their effects through exudation, exfoliation, and symbiosis (e.g., mycorrhiza) are also critically important. A great deal of primary methods development remains to be done in this area.

6. Some resolution is required on the importance of spatial scales in considering the effects of organisms on SOM dynamics. Organisms influence SOM on spatial scales from the individual aggregate to the landscape unit; the relevant scale of investigation must be kept in mind.

7. The potential of stable isototopes as tools for evaluating the effects of organisms on SOM pools and transformations is virtually unlimited. In particular, the influence of soil organisms in stabilizing or destabilizing SOM can be investigated in this way. Both labeling studies with multiple isotopes (^{13}C, ^{15}N, ^{34}S) and the exploitation of transitions from C_3 to C_4 vegetation (or vice versa) should be useful.

8. Where possible, standardized methods should be used in studies to ensure comparability of data. The Tropical Soil Biology and Fertility Handbook of Methods (J. M. Anderson and Ingram, 1989) provides a framework for standard methods for network studies on tropical soil processes.

Chapter 5

Organic Input Management in Tropical Agroecosystems

Pedro A. Sanchez, Cheryl A. Palm, Lawrence T. Szott,
Elvira Cuevas, and Rattan Lal

with James H. Fownes, Paul Hendrix, Haruyoshi Ikawa,
Scott Jones, Meine van Noordwijk, and Goro Uehara

Abstract

Organic inputs to the soil must be considered separate from soil organic matter (SOM) to understand properly the physical, chemical, and biological processes involved in the sustainability of tropical agroecosystems. Present management of organic inputs is largely empirical and qualitative. A better understanding of the role of organic inputs on the processes involved in litter decomposition, nutrient cycling, and soil aggregation is needed. The management of organic inputs should be as predictable as that of inorganic inputs, such as fertilizers and lime. The application of organic inputs should promote the synchrony of nutrient release with plant growth demands. Predictive parameters are needed to approach synchrony. Research on decomposition of organic inputs should be conducted in various agroecosystems. Very little is known about the role of root litter in tropical agroecosystems. Six research imperatives are identified for tropical agroecosystems: (1) quantify the biomass and nutrient content of aboveground organic inputs; (2) develop predictive parameters for nutrient release patterns, or quality, of organic inputs; (3) investigate the effect of placement of organic inputs on nutrient availability; (4) quantify production and nutrient release by roots, identify resource quality parameters (see Chapter 4) that describe root decomposition, and determine the relative importance of roots as sources of nutrients; (5) investigate the effects of quality and placement of organic inputs on soil physical properties; and (6) understand how organic inputs are transformed into functional SOM pools in soils differing in texture, mineralogy, and moisture and temperature regimes. We need reliable field methods for measuring root production and decomposition, transfer from organic inputs into SOM fractions, and transfer processes among SOM pools. The diversity of tropical species and farming systems provides ample combinations of crops and organic inputs for improving tropical agroecosystems.

Agroecosystems differ from natural ecosystems in that large amounts of biomass and nutrients are removed as crops. Nutrient outputs via crop harvests far exceed the other nutrient loss pathways combined. This is shown in Table 1 where a summary of annual crop nutrient budget studies from temperate-region countries is presented. A long-standing principle of sound agriculture is to replace the nutrients lost via crop harvests by adding either inorganic or organic fertilizers. Aside from adding nutrients, organic additions improve soil physical properties and help maintain soil organic matter content, all of which positively affect plant productivity (Allison, 1973). Organic inputs to the soil consist of above- and belowground litter, crop residues, mulches, green manures, animal manures, and sewage. Some of these inputs are grown and recycled on site

Table 1. Summary of nutrient inputs and outputs (kg ha^{-1} yr^{-1}) of 18 crop production systems from the United States, United Kingdom, Netherlands, France, Israel, and Japan. (Mean and standard deviations calculated from data of Frissel, 1978.)

Nutrient inputs or outputs	Nitrogen	Phosphorus	Potassium
Inputs			
Inorganic fertilizer	156 ± 189	39 ± 38	119 ± 188
N fixation	19 ± 39		
Other	13 ± 13	0.4 ± 0.8	9 ± 13
Total	188 ± 75	39 ± 39	127 ± 176
Outputs			
Harvest removal	103 ± 86	16 ± 16	91 ± 13
Denitrification	20 ± 24a	—	—
Leaching	3 ± 8b	nd	nd
Total	127 ± 95	16 ± 6	91 ± 13
Balance	60 ± 93	24 ± 30	37 ± 77

a. Mean of 10 out of 18 data sets reporting denitrification.
b. Mean of 3 out of 18 data sets reporting leaching.
nd = not determined.

Dynamics of SOM in Tropical Ecosystems

(internal to the system), while others are brought in from other sites, thus constituting an external source of nutrients and organic C.

Organic inputs are distinct from soil organic matter (SOM), which can be defined as organic material of biological origin that has undergone partial or complete transformation in the soil and is located below the soil surface. In much of the literature, organic inputs and SOM are lumped together as "organic matter," creating considerable confusion when attempting to evaluate and manage the role of organic inputs compared to that of soil organic matter. We propose, therefore, to discourage using the term *organic matter* and encourage the use of *organic inputs* and *soil organic matter (SOM)*.

Although early agriculture depended heavily on organic inputs, the emphasis shifted to inorganic fertilizers as they became abundant and economically practical. Today, because agricultural development in much of the tropics is largely limited by economic constraints, emphasis must again be placed on use of organic inputs along with chemical fertilizers. This paper briefly reviews

Figure 1. Conceptual model of major pools and transfers of soil organic matter. Numbers indicate focus of research imperatives (see text conclusions).

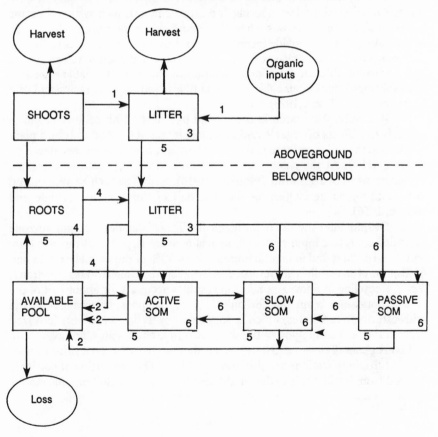

what is known about organic inputs in tropical ecosystems. We address the role of organic inputs on nutrient availability, soil physical properties, and soil organic matter content and composition. These issues are viewed in relation to plant production and how such information can be applied to the efficient management of organic inputs in tropical agroecosystems. A conceptual model is shown in Figure 1.

Historical Perspective

Continuous cultivation of food crops evolved first on flood plains of Mesopotamia and China on soils with ample reserves of weatherable minerals. Use of organic inputs was reported in the Middle East as far back as 2000 BC and later in continuous cropping systems in China, Japan, India, Mexico, Peru, and the Greek and Roman Empires (Allison, 1973). Although some inorganic fertilizers were sporadically used (ashes, lime, and potassium nitrate), continuous cultivation primarily depended on organic nutrient inputs until the beginning of the 20th century.

History also shows that when nutrients removed by crop harvest were not replaced by mineral weathering and organic inputs, such systems became unsustainable. One case was when the feudal system of Europe during the Middle Ages did not provide incentives for returning crop residues to the soil (Loomis, 1978). Another case occurred in the 1800s in the southeastern United States when insufficient use of organic inputs did not permit a stable transition from shifting to continuous cultivation, resulting in widespread erosion and land abandonment (Allison, 1973).

Agricultural research in the second half of the 19th century began to quantify the effects of organic and inorganic inputs. The need to return plant residues to the soil in order to maintain soil organic matter levels was identified (Lawes, 1889). At the same time, sharp crop yield responses to inorganic fertilizers were being shown (Hopkins, 1910). Such research revolutionized temperate-region agriculture in the first half of this century (Tisdale and Nelson, 1956).

Despite intensive use of inorganic fertilizers in modern temperate-zone agriculture, organic input returns remain at high levels, primarily through crop and root residues and animal manures. About 90% of the animal excreta and 68% of the crop residues are returned to the soil in the United States (Bertrand, 1983). Recycling of aboveground crop residues accounts for about half of the N and potassium input but for only 25% of the phosphorus input in the midwestern United States, which is dominated by fertile, intensively fertilized soils; in the acid soil region of the southeastern United States, however, crop yields are generally lower and the absolute amounts of both crop residue returns and fertilizer applications are also lower (Table 2). The proportion of nutrients derived from fertilizers is higher in the less fertile region. Soil organic matter

Table 2. Relative contribution (million kg yr^{-1}) of nutrient inputs from crop residues and fertilizer in two regions in the United States in 1979. (State averages [mean of six states for each region] calculted from data of Lindstrom and Holt, 1983.)

Nutrient input	Midwestern U.S. (Mollisol-dominant)			Southeastern U.S. (Ultisol-dominant)		
	N	P	K	N	P	K
Fertilizers	460	129	298	149	40	88
Crop residues	330	44	318	65	8	50
Total	790	173	616	214	48	138
Proportion of crop residues (%)	42	25	52	31	17	36

contents in some fertile, intensively fertilized Mollisols in the midwestern United States have stabilized or increased because of the large amounts of crop residues returned to the soil (W. P. Larson et al., 1978; Barber, 1979).

Our knowledge of the management of inorganic inputs has advanced to the point of quantitative prediction about the correction of plant nutrient deficiencies and the residual effect of fertilizer applications (Barber, 1984). A similar effort is needed for the management of organic inputs (Sanchez and Miller, 1986).

Organic inputs have been studied in detail but from two very different directions. One is a traditional approach in which organic residues are added and crop yields are reported. Unfortunately this approach has seldom been sufficiently rigorous or quantified to provide an understanding for observed differences. The other, a mechanistic approach, has been scientifically rigorous and has provided great detail about biological processes and chemical reactions involving organic additions in natural systems and laboratory experiments. This approach, however, has not provided much insight into the relevance of such processes to the function of agroecosystems. By unifying the two approaches, a predictive understanding can be developed, leading to the design of more efficient tropical agroecosystems.

Organic Inputs in Tropical Agroecosystems

Since very little is known about organic inputs in tropical agroecosystems, we will draw on principles derived from ecological and agronomic research in the temperate zone to guide research objectives. At the same time we must be aware of the differences between tropical and temperate ecosystems and within tropical ecosystems that may preclude extrapolation of results. For example, lowland humid tropical ecosystems have a continuous growing season, whereas temperate ecosystems do not. In the arid and semiarid tropics, the growing season also may be continuous with adequate water, as in irrigated areas. Rainfed agriculture in the seasonal or semiarid tropics, however, is subject to great spatial and temporal variation in rainfall and unpredictable spacing between rainfall events. These differences affect both the amount of biomass produced and the amount of organic material decomposed in a year, with important implications for the management of organic inputs in tropical agroecosystems.

Differences in predominant soils between climatic zones also have important implications for the management of organic inputs and soil organic matter. Much soil organic matter research has been conducted in the temperate glaciated soils, which are generally fertile, high in pH, and dominated by high-activity, permanent-charge clays. On the other hand, soils of the humid tropics are generally acid, nutrient-deficient, and dominated by low-activity or variable-charge clays (Chapter 3). Soils of the seasonal and semiarid tropics vary widely in these characteristics. These differences in soil conditions are reflected in differences in soil chemical, physical, and biological processes relevant to decomposition of organic inputs, formation and stability of soil organic matter, and nutrient availability.

Soil survey information and Soil Taxonomy (Soil Survey Staff, 1987) describe the distribution and characteristics of different soils and may provide an indication of biomass and nutrient content of above- and belowground inputs for various soil-vegetation systems. For example, taxonomic names or soil characterization data provide qualitative information on the decomposition state of the surface or near-surface litter and the amount and size of roots and pores, as well as quantitative data on chemical and physical properties of the soil.

Tropical agroecosystems also differ from those in temperate zones in that they offer more options for the management of organic inputs. Research in temperate-zone agriculture on type, quality, and quantity of organic inputs has generally focused on crop residues (Ochwald, 1978), with some work on cover crops. There has been less work on the importance of the timing of additions because of the break in growing season. In the tropics, particularly the humid tropics, there exists a greater diversity of agroecosystems. Among these agroecosystems, in addition to continuous cultivation, are short-term fallows, plantations with cover crops, tree-tree systems (such as coffee-*Inga*, or cacao-*Erythrina*) and tree-food crop systems (such as alleycropping). The tropics also house a diversity of plants that are used in agroecosystems and are virtually unknown to temperate-zone agriculture. There are indications that plants growing on acid soils, which include the majority of tropical soils, contain more secondary compounds (polyphenols) than those on neutral soils (McVey et al., 1978; Muller et al., 1987). These secondary compounds may affect decomposition and nutrient release and perhaps even soil organic matter formation.

Dynamics of SOM in Tropical Ecosystems

Farmers make choices about when, where, and how to add various organic inputs and decide on whether crop residues are used as fodder or as a mulch or are incorporated into the soil directly through plowing. In a number of tropical areas, notably the semiarid regions of Africa, nomadic people have started to grow crops in more or less fixed locations (Salzman, 1980). In such circumstances, the use of organic inputs has only recently developed and the quantity, quality, and seasonal variation of organic amendments, notably animal manure, is virtually unknown.

Because many farmers live in areas that are marginal for agriculture and subject to unreliable rainfall, the management of organic inputs in semiarid regions requires particularly careful study. Anthropologists working alongside agronomists and ecologists may help clarify the choices made by farmers for whom locally obtained organic inputs are presently the only option for improving soil fertility.

Nutrient Availability

One of the functions of organic inputs is to supply nutrients to crops. Nutrient release and availability depends on the rate of decomposition, which is controlled in part by temperature, moisture, soil texture, and mineralogy, as well as by the rate of application, placement, timing, and quality of the organic inputs. Much of the knowledge about the process of decomposition in agroecosystems has been obtained from soils dominated by permanent-charge minerals, usually well supplied with bases but limited by N and with microbial populations dominated by bacteria. It is likely that decomposition of organic inputs and subsequent nutrient availability will operate differently in acid, infertile Oxisols and Ultisols with variable-charge minerals, often high in phosphorus-fixation capacity, and with microbial populations presumably dominated by fungi. For example, Cuevas and Medina (1988) found that decomposition of forest litter on Amazonian Oxisols was limited by P, Ca, and Mg rather than by N. This leads to several questions about the controls on decomposition in tropical soils. Are N mineralization/immobilization patterns different in soils dominated by decomposer organisms with higher C/N ratios? Is the mineralization of organic phosphorus similar or different in such soils? In Andosols where microbial activity may be severely limited by P (Munevar and Wollum, 1977), are different ways of managing organic inputs needed? Much has been learned about the interactions between organic substances, microbes, and soil minerals (Huang and Schnitzer, 1986), but very little is known about the consequences of such interactions on plant production at the field level, not only in the tropics but anywhere.

In natural systems, nutrient release from litter and plant uptake of nutrients generally occurs in synchrony, resulting in efficient use of nutrients. In agroecosystems, the two processes of release and uptake are often separated in time, resulting in low nutrient-use efficiency. This is particularly acute with N, where excess amounts are lost by leaching, denitrification, and ammonia volatilization. There are few sound data on the magnitude of such losses (EPA, 1988; White, 1988), but in general only 30 to 50% of the N applied as fertilizer is recovered by most crops to which it is added. Although much of the unaccounted-for N is probably incorporated into SOM or roots, considerable losses do occur. N recovery from organic inputs does not appear to be better than that from inorganic fertilizers in the short term (Smith et al., 1987). However, most studies on N recovery from organic inputs only consider the first year; increases in efficiency may appear if residual, or long-term, effects are considered. An understanding of the processes that control nutrient availability from organic inputs may enable the management of more efficient tropical agroecosystems (Swift, 1985). From a management viewpoint, nutrient availability can be controlled to some extent by the quantity, quality, placement, and timing of organic inputs.

Nutrient Quantity

The amount of nutrients added via organic inputs depends on the mass of organic inputs and the concentration of nutrients in the tissues. Both the biomass and nutrient concentrations will vary with the soil properties, climate, and production system under which the organic input is grown. There is a wide body of knowledge about organic inputs and the nutrients they contain in natural tropical ecosystems such as rainforests (Vitousek and Sanford, 1986) and savannas (Sarmiento, 1984). In spite of abundant reviews, reliable estimates of biomass and nutrient inputs in tropical agroecosystems are few (see reviews by FAO 1975, 1978, 1985; Rosswall, 1980; Robertson et al., 1982; Wetselaar and Ganry, 1982; Cabala-Rosand, 1985; Kang and van der Heide, 1985; Wilson, 1988).

A rough estimate of the aboveground organic input potential of some agroecosystems is shown in Table 3 and compared to that of some natural tropical ecosystems. This table excludes belowground inputs. The following discussion refers to Table 3 and is based on the assumption that all crop residues remain on the soil, which is definitely not the case in many systems where residues are burned *in situ* or are used for fuel or as forage for domestic animals.

Natural Ecosystems

In the natural systems, tropical rainforests, both on acid and more fertile soils, have higher levels of aboveground litter inputs than do tropical savannas (see Table 3). The influence of native soil properties is evident, with lower biomass and nutrient inputs in acid Oxisols and Ultisols than in the more fertile Alfisols and Andosols (Vitousek and Sanford, 1986).

Tropical Food Crops

Carbon and nutrient inputs through aboveground residues for some of the most important grain crops of the tropics at current yield levels are generally less per crop than they are in natural systems (see Table 3). Part of the difference is due to the average crop growth duration of only 4 months rather than the entire year of litterfall in a humid natural system. The only root crop for which adequate data were available, potatoes in the Peruvian highlands, had C inputs as high as those of the natural ecosystems. Although important crops such as grain legumes and cassava are not included in this comparison, crop residue biomass production at present crop yield levels is clearly less than aboveground biomass production of tropical rainforest and savanna ecosystems. But seldom are crops grown only during 4 months of the year and the land is left fallow the rest of the time.

High-Input Systems

A more interesting comparison is between natural systems and tropical farming systems on an annual basis (see Table 3). Examples are divided into high- and low-input cropping systems, defined in terms of chemical and labor inputs (Sanchez and Salinas, 1981). The high-input agroecosystems generally have C and nutrient inputs similar to the natural systems. For example, returning residues in an annual maize-maize-soybean rotation in the Amazon of Peru provides as much C, more N, more P, more K, and similar levels of Ca and Mg than does the litterfall of tropical rainforests on similarly acid soils. Residue return in intensive, flooded rice production exceeds litterfall in tropical rainforests on fertile soils; potassium, in particular, is returned at much higher rates than in the natural system. A high-input agroforestry system on fertile soils of the humid tropics, cacao under *Erythrina* shade, recycles less C and N than the natural system on similar soils. Large areas of acid savannas of tropical America have been converted to intensive soybean production. Studies show that although only one crop can be grown per year, C and nutrient inputs from soybean residues are similar to those of the native savanna except for N and Ca, which far exceed levels in the natural system.

It is necessary to point out that all these high-input production systems are based on intensive use of inorganic fertilizers in addition to the recycling of organic residues. The general conclusion is that C and nutrient input from recycling crop residues in such tropical agroecosystems generally equal or exceed that of the natural systems. This conclusion is a direct result of the application of external sources of nutrients, in most cases inorganic fertilizer, which increases nutrient availability and results in higher plant production. In specific instances, high-input agroecosystems far exceed natural systems in one or more key nutrient inputs (see Table 3).

Table 3. Calculated biomass and nutrient content of aboveground organic inputs in some natural and agroecosystems in the tropics.

System description and duration (in parentheses)	Economic yield (Mg ha^{-1})	Yield/ residue ratio	Material	Biomass dry matter (Mg ha^{-1})	Element input (kg ha^{-1})					
					C	N	P	K	Ca	Mg
NATURAL ECOSYSTEMS (1 yr)										
Tropical rainforest, acid soils[b]	—	—	Litterfall	8.8	3960	108	3	22	53	17
Tropical rainforest, fertile soils[b]	—	—	Litterfall	10.5	4725	162	9	41	171	37
Tropical savanna, Venezuela[c]	—	—	Litterfall	3.8	1710	25	5	31	10	11
SINGLE CYCLE TROPICAL FOOD CROPS[d]										
Rice (4 months)[e]	2.2	0.8	Straw	2.8	1260	15	2	37	11	7
Maize (4 months)[f]	1.3	0.7	Stover	2.0	900	18	3	19	7	4
Sorghum (5 months)[f]	0.7	0.8	Stover	0.9	405	8	1	10	4	2
Potatoes (4 months)[f]	10.7	0.7	Tops	15.1	6795	17	1	43	14	7
Soybeans (3 months)[g]	1.7	0.7	Stover	2.4	1080	27	2	24	22	8
HIGH-INPUT TROPICAL AGROECOSYSTEMS										
Flooded rice, 2 crops per year (1 yr)[e]	11.0	1.0	Straw	11.0	4950	59	9	151	43	29
Maize-maize-soybean rotation (1 yr)[h]	8.7	—	Stover	9.3	4185	139	15	98	52	23
Soybeans in the Cerrado (4 mos)[g]	2.3	0.7	Stover	3.5	1575	86	8	43	32	11
Cacao/Erythrina in Brazil (1 yr)[d]	1.0	0.2	Leaf litter	6.0	2700	81	14	17	142	42

LOW-INPUT TROPICAL AGROECOSYSTEMS

Upland rice, rice-cowpea rotation, Peru (1 yr)[j]	4.7	—	Straw/stover	6.0	2700	77	12	188	27	12
Legume-based, (Brachiaria humidicola/ Desmodium ovulifolium) pasture, Colombia (1 yr)[l]	—	—	Leaf litter	7.0	3153	60	5	12	60	13
Alleycropping, Inga edulis (1 yr)[k]	—	—	Tree prunings	6.0	2700	137	10	52	32	—

a. Calculated as 0.45 x biomass dry matter.
b. Vitousek and Sanford (1986).
c. Sarmiento (1984).
d. Crop yields calculated from FAO (1985) for tropical countries.
e. Nutrient content calculated from DeDatta (1981).
f. Nutrient content calculated from Sanchez (1976).
g. Yields from Goedert (1986); nutrient content from Henderson and Kamprath (1970).
h. TropSoils (1987); Sanchez (1976); Sanchez et al. (1983).
i. CEPLAC (1985).
j. Sanchez and Benites (1987).
k. CIAT (1985).
l. Szott (1987).

Low-Input Systems

What is the case for low-input systems where fertilizers and lime are kept to a minimum or not applied at all and where the soil acidity constraint is taken care of by the use of aluminum-tolerant species? Examples are presented in Table 3. An upland rice-cowpea rotation grown after burning a secondary forest without fertilizer additions recycled less C and N but more P and much more K than did the rainforest on similarly acid soils. A low-input, acid-tolerant *Brachiaria humidicola/Desmodium ovalifolium* pasture in Oxisols of the Colombian Llanos returned more C, N, and Ca, similar levels of P and Mg, but less K to the soil than did the tropical savanna ecosystem.

Many low-input agroforestry systems have the potential to equal natural systems in terms of C and nutrient inputs. Szott (1987) showed that an alleycropping system on Ultisols with *Inga edulis* as the leguminous hedgerow produces 6 Mg ha^{-1} yr^{-1} of dry matter prunings containing similar amounts of N and P, higher K, but lower Ca than the natural forest (see Table 3).

Lack of Data

The dry matter and nutrient contents of animal manures used as organic inputs are seldom known. Analysis of the extensive body of literature comparing organic manures with inorganic fertilizers in the tropics shows that results depend largely on the nutrient content of the manures (Sanchez, 1976). In dry tropical Africa, for instance, agropastoralists typically move their animals from place to place according to forage availability. For much of the year, manure is left where it falls and its deliberate incorporation into soils where crops are grown may be confined to the postharvest period when livestock graze on crop residues.

Nutrient contents of organic inputs other than N are seldom reported, but they are extremely important in low-fertility soils. Information on the P, K, S, and micronutrient contents of crop residues and animal manures is scarce. In cereal crops most of the Zn, Cu, Mn, Fe taken up by the plants end up as aboveground crop residues, with some exceptions (Table 4). Sulfur is highly

Table 4. Proportion of nutrients (% in crop residue) accumulated by cereal crops present in aboveground residues. Root residues are ignored. (Data calculated from Sanchez, 1976; DeDatta, 1981.)

Crop	C	N	P	K	Ca	Mg	S	Zn	Cu	Mn	Fe
Rice	56	30	26	80	87	68	38	69	34	91	83
Maize	61	43	35	77	98	70	57	68	57	96	95
Sorghum	56	54	43	86	95	77	1	55	63	78	83

variable. Since many tropical soils are marginal or deficient in S and micronutrients, data about these nutrients are obviously needed.

The first research imperative, therefore, is to quantify the biomass and nutrient content of aboveground organic inputs in tropical agroecosystems. A solid, scientifically rigorous data base obtained on well-characterized soils is the necessary first step toward a better understanding of the processes involved in the management of organic inputs.

Quality of Inputs and Timing of Application

The time at which nutrients are made available to plants is as important as nutrient quantity. In farming systems where trees or shrubs are pruned for green manure, both the total quantity and the ratio of leaves to wood in the organic inputs can be manipulated by frequency and intensity of cutting. The efficiency of nutrient transfer from organic inputs to crops might be managed by varying the quality or the timing of application of organic inputs (Swift, 1985).

Quality of crop residues was originally defined by the C/N ratio (Jensen, 1929). Materials with C/N less than 20 generally decompose quickly and mineralize N immediately, whereas those with wider ratios decompose slowly and even immobilize N. It was also recognized that the type of C compound influenced the rates of decomposition (Tenney and Waksman, 1929). The lignin to N ratio proved to be a better predictor than C/N ratio for leaf litter decomposition in a variety of temperate forest species (Melillo et al., 1982). Recent work suggests that N release from legumes (low C/N, low lignin/N) is better correlated with polyphenolic content than with lignin and N content (Vallis and Jones, 1973; Palm, 1988). Legumes with high N content but different N release patterns have interesting implications for organic input management and is an area that merits further research.

Much has been said about the effect of the quality of organic inputs on N release and availability. Little is known, however, about quality with respect to P availability. This "nitrogen bias" is a result of the N-limitation in many temperate ecosystems, coupled with the difficulty of measuring P availability and mineralization, particularly in high P-fixing soils. Research priority must be given to the management of phosphorus via organic inputs because P often will be the ultimate limitation to crop productivity, especially in the tropics where high P-fixing soils are abundant and where the use of leguminous plants as organic inputs can often solve N limitations. Furthermore, crop residues are usually low in P, because most of this element is concentrated in the grain and is removed by harvest. The quality of organic inputs could affect P availability through (1) the P content or the C-to-P or N-to-P ratio of the material, (2) the size and activity of the soil microbial pool (Hedley et al., 1982), and (3) the interactions between the organic material and the mineral soil.

Some organic inputs, when incorporated into the soil, temporarily reduce aluminum toxicity. The process involved is believed to be a complexation of aluminum in the soil solution by organic acids, polysaccharides, and other initial products of organic input decomposition (Hue et al., 1986). The effect may be temporary because as these products undergo further decomposition, aluminum is released again to the soil solution. Little is known about the influence of organic input quality on aluminum complexation. Are high-quality inputs more effective than low-quality? If they are, this effect, although temporary, may be of agronomic significance when green manures are incorporated on acid soils in low-input systems. Another factor to consider is the reduction of aluminum saturation by the Ca and Mg released from the organic inputs (Wade and Sanchez, 1983).

Criteria that predict nutrient-release patterns will help in the selection and management of organic inputs for a diversity of agroecosystems. Once mineralization/immobilization patterns are known for a variety of plant types (based on some chemical criteria), agroecosystems can be managed more efficiently by selection of organic inputs with nutrient immobilization-release patterns more similar to the nutrient uptake patterns of specific crops. This synchrony of nutrient release and uptake also depends on the appropriate timing of application. A second research imperative, therefore, is to develop predictive parameters for nutrient release patterns, or quality, of organic inputs to tropical agroecosystems.

Emphasis should be placed on characterizing legumes and also on defining quality in terms of phosphorus availability. In the semiarid tropics, where biomass production varies from year to year and decomposition rates are episodic and variable, the development of predictive models is going to be correspondingly more difficult.

Placement of Organic Inputs

The placement of organic additions affects the physical and biological properties of the soil, which has important consequences for the temporal and spatial availability of nutrients and the potential for nutrient loss (see review by Doran and Smith, 1987).

The biological environment for decomposition is quite different for organic inputs left as mulches on the soil surface than for those incorporated into the soil by tillage (Parr and Papendick, 1978; Holland and Coleman, 1987). In general, macrofaunal activities and populations are greater in no-till systems (Lal, 1987; Pashanasi and Lavelle, 1987), and microbial populations may shift from predominantly fungal to bacterial with tillage (Holland and Coleman, 1987). Decomposition is generally faster for incorporated than for surface-applied material. However, much still needs to be known about the subsequent availability of nutrients, given the biological and physical environment in which decomposition takes place. Information is especially lacking for the humid tropics where biological activity is generally high but where, at the same time, temporary reducing conditions can occur in microsites of well-drained soils.

There are several questions of interest with respect to placement of organic inputs and nutrient availability. How do mineralization/immobilization patterns of organic inputs change with changes in macrofaunal or microfaunal populations that are caused by tillage (e.g., Hendrix et al., 1986)? Are N losses through ammonia volatilization due to surface placement of organic inputs significant in acid soils? Are losses of N through denitrification greater with surface application or incorporation of organic inputs in humid, tropical environments? Are these losses significant? What is the interaction between the quality of organic inputs and placement with respect to nutrient availability and losses? How does P availability differ with tillage practice? Answers to questions such as these will help in improving nutrient transfer from organic inputs to crops. A third research imperative, then, is to investigate the effect of placement of organic inputs on nutrient availability in terms of the interaction between soil temperature, moisture, redox potential, soil fauna, and input quality.

Roots as Organic Inputs

Although there is limited information about the nutrient content and quality of aboveground inputs, virtually nothing is known about the role of roots as an organic input. Roots are generally not considered in nutrient cycling studies because of difficulties in measurement. In tropical and temperate natural forest ecosystems, they are known to play a major and often dominant role in nutrient cycling (Fogel, 1980; Cuevas and Medina, 1986). In agroecosystems, fine-root decomposition is likely to be an important source of nutrients because it occurs within the crop rooting zone.

There has been considerable work carried out on root production, turnover, and nutrient release in temperate forests (Safford, 1974; Santantonio et al., 1977; Persson, 1980, 1983; Vogt et al., 1980, 1982; Keyes and Grier, 1981; McClaugherty et al., 1982), but very little is known about root turnover and nutrient release in tropical ecosystems (Jordan and Escalante, 1980; Cuevas and Medina, 1983; Sanford, 1985; Hairiah and van Noordwijk, 1986). In studies carried out in the temperate zone, Fogel (1980) found that in a *Pinus taeda* plantation, the fine root fraction contained 84% of the N in aboveground litterfall but released twice the N during decomposition. Vogt et al. (1982) estimated that fine roots plus mycorrhiza in a stand of *Abies amabilis* cycle about four times the N, six to ten times the P and K, two to three time the Ca, and three to ten times the Mg than that contained in litterfall.

In an Amazon forest on an Oxisol in Venezuela, Cuevas (1983) found that annual fine-root production in the first 10 cm of soil and the root mat above the mineral soil, was $8 \, Mg \, ha^{-1} \, yr^{-1}$. Sanford (1985) found that fine root turnover in the upper 10 cm of the same Oxisol was 25% per month. N and phosphorus concentrations of these roots were 2.3% N and 0.11% P, so the quantities of N and P added to the soil by fine root turnover were very large – 190 kg N ha^{-1} yr^{-1} and 9 kg P ha^{-1} yr^{-1}, respectively – assuming no retranslocation of nutrients from the roots before death. Biomass and nutrient fluxes in litterfall for the same forest amounted to 7.6 Mg dry matter ha^{-1} yr^{-1}, 121 kg N ha^{-1} yr^{-1}, and 2 kg P ha^{-1} yr^{-1} (Cuevas and Medina, 1986), indicating that fine-root production and turnover in forests on very infertile soils may be more important for nutrient cycling than litterfall.

Carbon and nutrient inputs from roots in tropical agroecosystems are even more poorly quantified. When determined in a low-input upland rice-cowpea rotation, roots (belowground residue) constituted 11 to 22% of the dry matter and from 5 to 21% of the nutrients accumulated by these crops (Table 5). R. J. Scholes and A. Salazar (personal communication) measured differences in root production among several crops in an Ultisol of Peru. Their estimates indicate that root litter inputs may be on the order of 1-4 Mg ha^{-1} per crop and higher in fertilized systems where crop production is greater; approximately one-third to one-half of the organic inputs to the soil originate from dead roots (Table 6). Assuming two to three crops per year, then, root production in tropical agroecosystems is similar in magnitude to that estimated for tropical forests (Jordan and Escalante, 1980; Cuevas, 1983; Vogt et al., 1986; Berish and Ewel, 1988; R. J. Scholes and A. Salazar, personal communication).

Such estimates, however, are based on root biomass measured near crop harvest; therefore they do not include C and nutrient inputs lost to soil via root turnover, sloughing, or exudation during the crop cycle. Exudates and exfoliates can account for a high proportion of total C transfers belowground (Martin, 1977; Milchunas et al., 1985) and may be more readily decomposable than root litter. Roots may further increase nutrient availability by producing organic acids, surface phosphates, and siderophores (Bowen, 1984).

Table 5. Partitioning of nutrients accumulated (%) by upland rice and cowpea crops in a low-input system in the Amazon. Mean of five upland rice crops and two cowpea crops. (Data from Sanchez and Benites, 1987; R. J. Scholes and A. Salazar, personal communication.)

| Crop | Component | Dry matter | Nutrient | | | | |
			N	P	K	Ca	Mg
Upland rice	Harvest removal	41	46	67	9	7	32
	Aboveground residue	48	43	25	85	79	63
	Belowground residue	11	11	8	6	14	5
Cowpea	Harvest removal	29	46	51	17	3	29
	Aboveground residue	49	42	35	77	76	52
	Belowground residue	22	12	14	6	21	19

Dynamics of SOM in Tropical Ecosystems

Table 6. Fine-root biomass (ash-free) to a depth of 30 cm, aboveground residues, and grain yields (all in Mg ha^{-1}) of various crops in Ultisols at Yurimaguas, Peru (standard errors for 18 samples in parentheses). Root litter production is assumed to equal fine-root biomass in these annual crops. (Data from R. A. Scholes and A. Salazar, personal communication.)

Plant	Fine-root biomass	Aboveground residue	Grain yield
Soybean	1.39 (0.42)	2.76	1.80
Cowpea	1.03 (0.09)	1.70	0.86
Maize	0.97 (0.12)	3.03	3.00
Upland rice	0.62 (0.12)	1.60	1.41

Temporal and Spatial Distribution of Root Litter

A number of questions about the importance of roots as a source of nutrients in tropical agroecosystems relate to temporal and spatial distribution of root litter and root litter quality.

Although annual root litter inputs may be similar in forests and agricultural fields, their temporal distribution is likely to vary between ecosystems. It contrast to forests, where inputs may occur throughout the year or be concentrated during dry periods, most of the root litter inputs in monocultural agroecosystems is believed to occur close to crop harvests. Greater proportional losses of added C and nutrients may occur in agricultural systems than in forests due to peaks in litter inputs in the former. Ladd et al. (1983) showed that retention of organic residues in soils was greater when smaller amounts of substrates were added. These differences would probably be reduced in multiple-cropping systems where harvests are staggered throughout the year or in agroforestry systems where perennial vegetation is present. The magnitude of the effect of the timing of root litter input on soil organic matter and nutrient release in tropical agroecosystems needs further study. Such results may aid the selection of species and their management over time.

The spatial distribution of root litter can be managed by choice of crops having different rooting patterns and by the use of tillage. It is commonly assumed that in agroecosystems based on short-cycle food crops the majority of the roots and root litter is found in the top 20 cm of soil, whereas in systems based on perennial crops more roots are found at greater soil depths. This may or may not be true. Tillage tends to homogenize root litter within a relatively narrow depth interval.

The spatial distribution of root litter also varies with the successional stage of the vegetation. Root distribution follows nutrient distribution in both space and time. In early successional stages when the litter layer is not yet formed, fine roots extend deeper into the soil (Berish, 1982; Bowen and Nambiar, 1984). As the litter layer develops with succession, nutrient distribution is more concentrated near the soil surface and more fine roots are found in the superficial layers. This concept may have important implications in the selection of trees for managed fallows or alleycropping.

Of special relevance here is the possibility of a continuous channel system into the subsoil left by decaying roots of a previous crop or vegetation, provided soil tillage does not disturb continuity. Roots from a cut-over forest may be important for crop production in shifting cultivation because crops are often planted close to old tree stumps. Several deep rooted cover crops (*Centrosema* and *Crotolaria* species, for example) may contribute to subsequent crop growth this way as well. The quantity of organic residue left by the roots is less important than the physical quality in modifying the soil pore system. A shallow coating of organic inputs around the root may help to overcome subsoil aluminum toxicity (Hairiah and van Noordwijk, 1986).

Many of the spatial effects of root litter production on soil physical and biological properties are well known, but a number of intriguing questions regarding the spatial distribution of root litter/nutrient-related processes remain. For example, what is the absolute and relative production of root litter at depths greater than 20 cm in various types of agroecosystems? How do rates of litter production and soil properties change with depth, and how do they interact in determining rates of mineralization/immobilization, denitrification, and phosphorus fixation? How does spatial heterogeneity of root litter affect rates of nutrient mineralization/immobilization? Can techniques other than tillage and crop species selection be used to manipulate use efficiency? For example, pruning of woody hedges in alleycropping systems affects the temporal and spatial distribution of hedge root litter and hence nutrient release to the associated crop.

Root Litter Quality

There has been no systematic comparison of the quality of root litter from agricultural crops, perennial crops, and tropical forests. Root litter from forest and crops is likely to differ in degree of homogeneity, proportion of fine roots, quantity of secondary or allelopathic compounds, nutrient concentrations, and the amount of lignin and polyphenols present. In general, small-diameter roots, such as those produced by food crops, have comparatively low lignin/N ratios and would be expected to decompose and release nutrients rapidly, whereas decomposition and nutrient release from larger, more lignified root litter would be slower (Amato et al., 1987; Berg et al., 1987). These indices of quality, however, may not be good predictors of root decomposition (McClaugherty et al., 1982); other indices, such as the nonstructural carbohydrates/N ratio, may be better indications of root litter quality (Berg et al., 1987).

Most work related to root decomposition uses fresh roots incubated in mesh bags. Under such conditions, intimate association between roots and soil may not exist; in addition, decomposition rates of fresh material may be different from those of dead roots since nutrient retranslocation from senescing roots is likely to affect quality. Key experiments measuring retranslocation from roots under varying environmental conditions have yet to be performed. Furthermore, the presence of allelopathic compounds may, at times, override the usual determinants of litter quality. Better understanding of the relationship between site fertility and root litter quality, and, indeed, the definition of a parameter that can be used to accurately predict decomposition and nutrient release from root litter are needed. Moreover, controls on root productivity and mortality are still in question (Caldwell, 1979; Marshall and Waring, 1985, Nadelhoffer et al., 1985).

Synchronizing root turnover and decomposition to the benefit of crop plants requires knowledge of the magnitude, spatial distribution, and time course of root production, mortality, and decomposition. Because of methodological difficulties, however, these estimates are still rather crude.

Estimates of total root production during a growing season have to be based on quantification of biomass at one point in time and estimates of root turnover. Techniques for measuring root production based on frequent sampling are subject to serious methodological criticism (Singh et al., 1984). Recently an alternative approach has become available by sequential observations of individual roots behind transparent walls (minirhizotrons) or in holes filled up with inflatable structures, between observations. Simultaneous root consumption by soil fauna, root decay, and root growth can now be recorded (van Noordwijk, 1987; Taylor, 1987). Results so far show turnover varies greatly between crops and soil conditions. Application of these techniques to tropical forests and agroecosystems is badly needed.

In summary, then, a fourth research imperative is to quantify root production and nutrient release by roots in tropical agroecosystems, to identify quality parameters that describe decomposition and nutrient release from roots, and to determine the relative importance of roots as a source of nutrients. Regardless of the answers to the above questions, perhaps the more relevant question with respect to roots is, can they be managed?

Organic Inputs and Soil Physical Properties

This discussion has been limited so far to the role of organic inputs as a source of plant nutrients. In addition to managing nutrient-release patterns from organic inputs, it is possible to manipulate other soil factors to maximize plant growth and nutrient uptake. Organic inputs may be used in two ways to manipulate the physical behavior of soils. They may be used as surface mulches or incorporated into the soil as an amendment.

Organic Inputs as Mulches

In situations where high surface temperatures, rapid dehydration of the seedbed and formations of dense, impervious surface crusts from raindrop impact are serious constraints to seedling emergence and survival, the use of organic inputs as mulches far outweighs the benefits of incorporating them into the soil for their amendment effects.

Mulch mainly regulates soil temperature (Maurya and Lal, 1981) and moisture regime (Figure 2). It buffers and dampens the effects of environmental factors on soils: it lessens the extremes, lowers the temperature maxima in hot and dry conditions, and raises the minima in cold and moist environments (Lal, 1979, 1987). The dampening effect is related to the amount of radiation penetrating the soil surface. The latter is influenced by mulch through radiation absorption, alteration of albedo, and by its physical barrier. While albedo is a function of mulch color, the barrier effect depends on the percentage of ground cover and the thickness of the mulch layer. The albedo is higher from fresh straw mulch than from decayed crop residue.

For the same rate of application, surface application of organic inputs has more drastic effects on soil temperature and moisture regimes than incorporation. Lawson and Lal (1979) studied the effects of rate and method of application of organic inputs on properties of an Alfisol at Ibadan, Nigeria. Mean maximum soil temperature decreased with surface application; with incorporation, mean maximum soil temperature increased with increasing mulch rate (Figure 3). Soil moisture content was also influenced by mulch rate and mode of application (Figure 4). Although the benefits of mulching for moisture retention and temperature regulation are obvious for soils of the seasonal tropics, little is known about such benefits in the humid tropics where water shortages are generally not a problem. Mulching under such conditions may have more negative (pests and diseases) than beneficial effects.

Mulch effects on some soil physical properties are related to activities of soil organisms. For example, surface mulching increases feeding and burrowing of earthworms and termites, which then alters bulk density, macroporosity, water retention, and transmission characteristics. In addition, increased production of earthworm casts and arthropod feces may increase the stability of aggregates (Rusek, 1986). Lal et al. (1980) observed that the activity of earthworm *Hyperiodrilus africanus* was linearly related to the mulch rate.

In sandy Alfisols of the semiarid tropics and subtropics, formation of dense, impervious surface crusts caused by raindrop impact on freshly tilled and seeded fields prevent seeds from emerging and subsequently increases runoff and soil loss. Work conducted in Niger on sandy Alfisols, where surface sealing is a major problem, showed marked improvement in water infiltration

Figure 2. Effects of mulch materials and tillage methods on soil moisture reserves in the Masika (rainy) season in Zanzibar, 1980. (Source: Khatibu et al., 1984.)

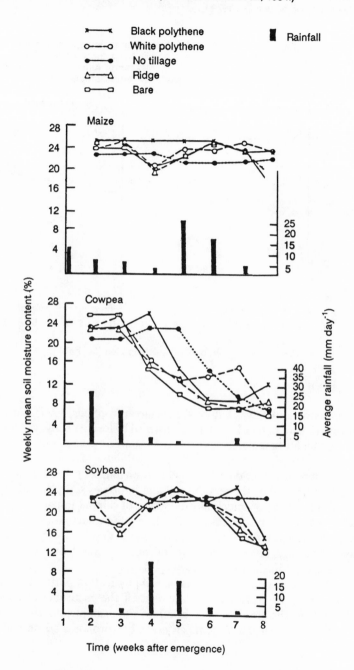

Figure 3. Relationship between average maximum and minimum soil temperature in the first two weeks after planting, amount of mulch applied, and method of application, 1974. (Source: Lawson and Lal, 1979.)

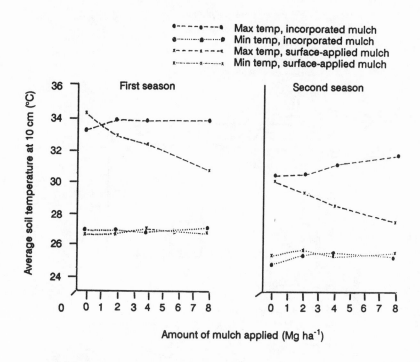

when a woody mulch was left on the surface (TropSoils, 1987). This resulted from an extensive network of channels formed by termite activity. Although the effects are highly beneficial, this practice is limited by the availability of organic inputs.

Mulch invariably decreases soil erosion and often reduces water runoff (Lal, 1984) because it shelters the soil and improves soil physical properties. Some of the effects of mulch are reduction of rainfall impact, improved soil structure and porosity, increased infiltration rate by decreased surface sealing, increased water retention capacity, decreased runoff velocity, and enhanced biological activity. For most soils, runoff and erosion decrease exponentially with increasing mulch rate (Lal, 1976).

The effects of organic inputs on soil physical properties depend on the residence time or quality and placement of the material. The longer the residence time, the more durable the effect. For example, the effects of mulching by rice and maize stover mulch are more pronounced on soil physical

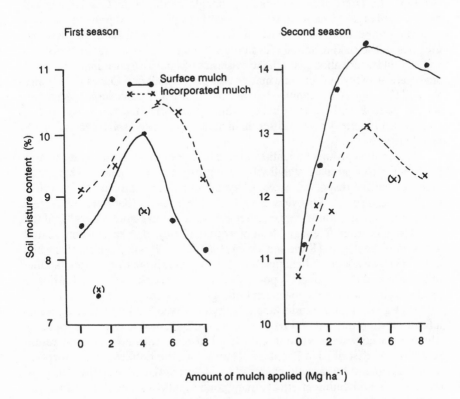

Figure 4. Relationship between mean soil moisture content, amount of mulch applied, and method of application, 1973. (Source: Lawson and Lal, 1979.)

properties than are those of high-quality residues such as cowpea or soybean because cowpea and soybean decompose quickly and have only transient effects on soil structure and water transmission characteristics.

Organic Inputs as Soil Amendments

The value of organic inputs as amendments to improve soil physical conditions may far outweigh their value as nutrient sources. Soils well supplied with organic inputs are generally looser, easier to work, more permeable to water, and less susceptible to drought; they possess a higher capacity to retain nutrients. Measurable effects of organic inputs on soil physical properties are commonly observed on total porosity and macroporosity, percentage of water-stable aggregates, and aggregate size distribution. Lal et al. (1980) observed an increase in aggregate size from 0.25-0.5 mm for plots with no inputs to 4-8 mm for plots with organic inputs at the high rate of 12 Mg ha^{-1}.

While there is widespread agreement that organic inputs improve soil physical conditions, there is less agreement on how organic inputs produce these effects. There are two schools of thought on this issue. One large group believes soil organic matter acts as a cementing agent to bind primary particles into compound units called peds or aggregates. The resistance of these aggregates to the slaking action of water is crucial to the structural integrity of a soil. The literature abounds with the existence of a positive correlation between aggregate stability and organic matter content. Tisdall and Oades (1982) also showed that aggregate stability depends on quality of organic inputs. Aggregation increased with lower quality inputs in poorly structured Alfisols of Australia. This effect may be different in better aggregated soils such as Oxisols and Andosols.

A smaller group of scientists adhere to the notion that soil organic matter contributes to aggregate stability by acting as a water repellent. This group believes that the dispersive power of water is rendered less harmful in water-repellent aggregates. This view is supported by the fact that field soils rarely achieve complete saturation, i.e., even at zero water pressure only 80% of the pore space is water-filled, and heat of wetting of clays decreases when treated with organic materials (Kijne and Taylor, 1964; Kijne, 1968). It is clear that while much is known about the beneficial effects of organic inputs, our understanding of the cause of these effects is poor. This lack of understanding is a barrier to predict and control the management of organic inputs.

Organic inputs render some soils less droughty by increasing water infiltration into the soil and by increasing the soil's water-retention capacity. This increased water retention goes hand in hand with an increased cation retention. As described in Chapter 3 (Oades et al., this publication), adsorption of organic matter by variable-charge clays increases the net negative charge of the clay. The amendment effects or organic inputs are therefore principally active in rendering the soil easier to work, more open, and accessible to oxygen and water and in enlarging the soil's capacity to retain and release water and nutrients to organisms that live in and on it.

Managing physical properties by selection and placement of organic inputs may be more important for some areas of the tropics than management for nutrients. As examples, management for erosion purposes will be particularly important for highly erodible soils such as soils with sandy A horizons. Management of organic inputs that enhance soil properties related to increased water capture and retention will be particularly important for the seasonal and semiarid tropics, whereas elimination of excess water could be important to plant growth in the humid tropics. A fifth research imperative, then, is to investigate the effect of the quality and placement of organic inputs and their interaction on soil physical properties such as microclimate and aggregation. Particular emphasis should be placed on the duration of the effects.

Organic Inputs and Soil Organic Matter

In addition to supplying nutrients and improving soil physical properties, organic inputs can also lead to the formation and maintenance of soil organic matter. In fact, many of the effects of organic inputs already discussed (e.g., nutrient availability patterns and aggregation) are indirect via the formation of soil organic matter. Numerous studies in both temperate and tropical regions have shown that soil organic matter can be maintained by additions of organic residues (Allison, 1973; Larson et al., 1978; Barber, 1979; Lal and Kang, 1982). However, there is little correlation between total soil organic matter content and plant growth or crop productivity (Sanchez and Miller, 1986). Perhaps more important to plant productivity than total soil organic matter is the effect of organic inputs on various soil organic matter fractions.

Recent models suggest dividing soil organic matter into a series of pools of different turnover times (Jenkinson and Rayner, 1977; van Veen and Paul, 1981; Jenkinson et al., 1987; Parton et al., 1987). From the agronomic point of view, the functional pools proposed by Parton et al. (1987) are most attractive among the many models in the literature. The active SOM pool could play a major role in nutrient release and perhaps can be estimated by microbial biomass determinations; the slow SOM pool may play a major role in the stabilization of macroaggregates, while the passive SOM pool may be largely inert as a nutrient release source but may play a major role in the binding of primary soil particles.

Currently there are no methods for separating SOM into the abovementioned pools and, in fact, these pools may not exist as discrete units. The discussion that follows presumes that methods will be developed. In addition, the discussion is highly speculative about the relationships among the suggested SOM pools, soil aggregates, and various chemical fractions in the soil. It is intended to serve as a conceptual framework for exploring the formation, maintenance, and importance of SOM.

A key challenge is to understand the relative role of above- or belowground inputs on the different SOM pools, in different agroecosystems, on soils with contrasting textures, mineralogies, temperature, and moisture regimes. The formation and maintenance of the various pools are thought to be affected by the quantity and quality of organic inputs, management practice, soil texture, and mineralogy. Key questions are how manageable are the soil organic matter pools and how do they relate to plant productivity?

In general, organic inputs high in lignin or polyphenolics, defined as low quality, lead to a more rapid formation of stable forms of soil organic matter (slow and passive pools) than plant materials of high quality (DeHaan, 1976; Martin and Haider, 1980; Stott et al., 1983; Kelley and Stevenson, 1987). High-quality inputs, in turn, are associated with increased active pool and nutrient availability, at least in the short term.

Little, however, is known about the long-term differences in the effect of input quality on SOM fractions and nutrient availability. In the case of N, some will go directly to the microbial pool (active pool) and be readily available in the short term via immobilization/mineralization processes. Some of the N that goes through the microbial pool will be associated with decomposition-resistant microbial compounds and become part of the slow or passive pools. Some N will react directly with polyphenolics, lignins, and breakdown products of SOM and be less readily available (slow and passive pools). The relative proportions of each pool formed for given qualities of inputs, however, is unknown. Although much is known about the fate of phenolics, lignins, and breakdown products of SOM, as well as readily decomposable C compounds in short-term laboratory experiments (Martin and Haider, 1980; Azhar et al., 1986; Stevenson, 1986), there are no long-term field data on the fate and availability of the various compounds, their connection to the various SOM pools, and their relevance to productivity.

Placement, tillage practice, and macrofaunal activities can also influence maintenance of SOM and the relative proportions of the various SOM pools through their effects on aggregation. This assumes that SOM associated with micro- and macroaggregates has different turnover times and perhaps corresponds to passive and slow pools, respectively (Tisdall and Oades, 1982; Tiessen et al., 1984c; Elliott, 1986).

Macroaggregate stability is in part controlled by roots and fungi through production of transient and temporary organic binding agents, through physical entanglement of soil particles within fine roots and hyphae, and through compressive forces of growing roots (Martin, 1977; Tisdall and Oades, 1982; Monroe and Kladivko, 1987). Tillage may reduce macroaggregate stability directly, by disrupting aggregates and exposing protected SOM to microbial attack, and indirectly, by altering growth and distribution patterns of roots and fungi. In contrast, it is thought that microaggregation is controlled largely by soil mineralogy and the binding of primary clay particles with persistent organic compounds. Therefore microaggregates may be less influenced by management (Tisdall and Oades, 1982; Tiessen et al., 1984). Elliott, 1986; Ramsay et al. 1986). The limits between these two functionally different aggregate sizes are not yet established for most soils and are likely to vary considerably with texture and mineralogy.

Macroaggregate stability, and presumably the size of the slow pool, should increase in conditions that favor fungal, rather than bacterial, populations. Fungi predominate in the decomposition of low-quality materials (Swift et al., 1981) and of surface-applied organic inputs (as opposed to incorporated organic inputs) (Holland and Coleman, 1987). A relatively larger slow SOM pool might be expected with incorporation than with surface placement of organic inputs because of the larger contact surface between organics and clay particles with incorporation and hence the greater aggregation and physical

Dynamics of SOM in Tropical Ecosystems

protection. Lower total SOM levels, however, are often observed in plowed soils, possibly due to disruption of macroaggregates, to reduced importance of fungi, and to the prevalence of bacteria and other soil organisms with high metabolic activity (Andren and Langerlof, 1983; Ryszkowski, 1985; Hendrix et al., 1986; Holland and Coleman, 1987).

Studies on the effect of placement and quality of inputs on the relative sizes of the various SOM pools are relatively new. Information on the factors controlling aggregation might be used to investigate how organic inputs can be manipulated to manage SOM pools and subseqent plant production.

Recent data from research in the temperate zone suggest that crop productivity is related to the size of the microbial (active) SOM pool (Janzen, 1987). Not only the size but also the turnover rate and composition (bacterial, fungal, and C/N) of the active pool must be considered. It is interesting to speculate whether microbial populations involved in decomposition can be manipulated, through either the type or placement of plant material to produce more decomposition-resisitant (slow pool) or more labile (active pool) soil organic matter in order to synchronize nutrient availability and uptake for the agroecosystem in question. For example, if one were interested in increasing the slow SOM pool, the strategy would aim at increasing the proportion of fungi in the microbial populations since fungi appear to stimulate soil aggregate formation and maintenance and the production of slowly decomposing soil organic matter (Tisdall and Oades, 1982; He et al., 1988a). Much needs to be learned about the potential for manipulating the active and slow pools by the management of organic inputs in terms of matching nutrient availability patterns to plant demands.

Assuming the various SOM pools can be controlled by varying the quality and placement of organic inputs, a simple scenario shows the possible relevance to nutrient availability for agroecosystems differing in nutrient-demand patterns. Short-cycle food crops have high demand for nutrients over a short time. Generally these demands are met by applications of fertilizers. In the humid tropics, however, there are often large losses of these fertilizer nutrients by leaching because the supply of readily available nutrients exceeds plant demand. Application of high-quality organic inputs (e.g., certain legumes) could provide a more efficient use of nutrients by releasing nutrients quickly to short-cycle food crops through rapid turnover or active SOM, but, over a longer time period, releasing nutrients more in synchrony with plant demand than do inorganic fertilizers. Mature tree crops, on the other hand, have lower nutrient demands over similar time periods than do short-cycle crops. Use of inorganic fetilizers or even high-quality organic inputs will probably result in large losses of nutrients. Application of lower quality plant residues may be more appropriate. Through a build-up of soil organic N in the slow pool by additions of low-quality inputs, there would be a lower but more continued release of N, more in synchrony with nutrient-demand patterns of trees. This scenario is based on inferences from laboratory and field studies and studies of natural ecosystems. It remains to be seen if these concepts can be applied to the management of agroecosystems in the field.

The sixth imperative, therefore, is to develop an understanding of the role of the quality and management of organic inputs on the formation of different SOM pools in tropical agroecosystems, on soils differing in texture, mineralogy, moisture, and temperature regimes.

Conclusions and Recommendations

Management of organic inputs needs to be as quantitative and predictable as management of inorganic inputs in agricultural systems. The diversity of tropical species and farming systems provides ample combinations of crops and organic inputs for designing efficient agroecosystems. Not only must the nutrient contents of the organic inputs be known or reasonably estimated, but also a set of quality parameters needs to be established to predict the rates of release of nutrients during the decomposition of organic inputs and how these patterns change with management and under the influence of various groups of soil organisms. Notably lacking are field methods for measuring root production and decomposition, transfers from organic inputs into SOM fractions, and transfer processes between SOM pools. Related to the difficulty of measurement is the fact that current conceptual models are not homologous with measurable SOM fractions and hence are difficult to test rigorously.

We propose the following imperatives:

Theme Imperative: Management of organic inputs needs to be as quantitative and as predictable as management of inorganic inputs in agroecosystems.

Research Imperatives:
1. Quantify the biomass and nutrient content of aboveground organic inputs in tropical agroecosystems.
2. Develop predictive parameters for nutrient release patterns, or quality, of organic inputs to tropical agroecosystems.
3. Investigate the effect of placement of organic inputs on nutrient availability.
4. Quantify the amounts of nutrients released by roots, identify quality parameters that describe decomposition and nutrient release from roots, and determine the relative importance of roots as sources of nutrients.
5. Investigate the effects of quality and placement of organic inputs on soil physical properties.
6. Develop an understanding of the role of organic inputs on the formation of functional SOM pools in soils differing in texture, mineralogy, and moisture and temperature regimes.

Chapter 6

Modeling Soil Organic Matter Dynamics in Tropical Soils

William J. Parton, Robert L. Sanford, Pedro A. Sanchez, and John W. B. Stewart

with Torben A. Bonde, Dac Crosley, Hans van Veen, and Russell Yost

Abstract

Studies of soil organic matter (SOM) in temperate ecosystems have been the source of data used to develop soil organic matter models. Fewer data are available for tropical soils, and soil organic matter dynamics are less well understood. In this paper we use the CENTURY soil organic matter model to simulate SOM dynamics in a tropical forest, a tropical grassland, and a tropical agroecosystem. The results suggest that the overall model structure is adequate for tropical soils and suggest some modifications needed to more accurately simulate SOM dynamics, nutrient cycling, and plant production in tropical systems. Specific modifications of the model will depend on results of research on the following topics: (1) effect of soil texture, mineralogy, and Fe and Al chemistry on SOM formation and decomposition; (2) fractionation of SOM in an active, slow, and passive pool; (3) phosphorus cycling, particularly soil organic P; (4) effect of pH on SOM dynamics; (5) extent of leaching of organics, including N and P compounds in tropical soils of varying mineralogy and texture; and (6) specific processes, such as denitrification, that determine the N and S budget. Unfortunately, it is difficult to use existing models (without the proposed modifications) because the data needed to initialize some of the site variables in the models are very rare.

Soil organic matter dynamics have been extensively studied for temperate soil systems during the past fifty years. The results of these studies have been incorporated into a variety of soil organic matter models. These models have been very effective at simulating changes in soil organic matter (SOM) that result from different agricultural practices (Parton et al., 1983 and van Veen et al., 1981) used for temperate agricultural systems. Unfortunately, research on soil organic matter in the tropics has been much less extensive and no models have been developed with tropical soils in mind. The major objective of this paper will be to discuss the problems associated with adapting the CENTURY soil organic matter model to soil systems dominated by variable-charge clay minerals and to demonstrate the use of the CENTURY model for a tropical forest system, a tropical grassland system, and a tropical agroecosystem. We will note the changes to the CENTURY model needed to adapt the model for soils with variable change and will speculate about further improvements needed for the tropical SOM model. The CENTURY SOM model was selected for this exercise because it incorporates most of the recent improvements in our understanding about SOM dynamics and includes the interaction of C, N, P, and S in the soil plant system (Parton et al., 1987, 1988).

Traditionally, the differences between temperate and tropical soils have been attributed to a set of soil characteristics which are supposedly normal for large regions in the humid tropics and either exceptional or less extensively developed for temperate soils. These characteristics are a deeply weathered profile and a mineralogy different from most temperate soils, with kaolinitic clay minerals most frequently cited as typical of tropical soils. With a global data base available at the scale of 1:5 million, this simplistic, stereotypic situation does not apply. The above-mentioned characteristics are found in large areas of the humid temperate region, notably southeastern U.S. and southeastern China. It is perhaps better to differentiate soils according to their intrinsic charge characteristics. Soils with mainly permanent-charge minerals occupy 10% of the tropics and 45% of the temperate region, while soils with predominantly variable-charge minerals (kaolinite, iron oxides, and allophane) cover 60% of the tropics but also 10% of the temperate region (Uehara and Gillman, 1981). The balance are soils with mixed charge.

It was commonly believed that soils of the tropics have lower organic matter contents than soils of the temperate region. The red color of many soils, high temperature, and high rainfall are among the reasons cited in support of this generalization. This assumption, however, no longer holds. There are no

Dynamics of SOM in Tropical Ecosystems

Table 1. Mean organic matter contents in 61 soils from the tropics and in 45 soils from in 45 soils from the temperate region (Source: Sanchez et al., 1982.)

Parameter	Depth (cm)	Tropical soils	Temperate soils	Significance	CVC (%) Tropics	CVC (%) Temperate region
Total C (%)	0-15	1.68	1.64	ns	53	64
	0-50	1.10	1.03	ns	57	69
	0-100	0.69	0.62	ns	59	75
Total N (%)	0-15	0.153	0.123	*	62	57
	0-50	0.109	0.090	ns	57	57
	0-100	0.078	0.060	**	54	52
C/N ratio	0-15	13.7	13.6	ns	79	35
	0-50	11.3	11.3	ns	46	32
	0-100	9.6	10.0	ns	46	35

$*P = 0.05$
$**P = 0.01$

major differences in SOM contents between the two regions (Sanchez and Buol, 1975; Lathwell and Bouldin, 1981; Post et al., 1985). For example, data from 61 randomly chosen profiles from the tropics and 45 from the temperate region classified as Oxisols, Mollisols, Alfisols, and Ultisols, provide no significant differences in total C and C/N ratios between soils from tropical or temperate regions at depth intervals up to 100 cm (Table 1).

Total N contents, however, were significantly higher in the tropical samples while the coefficients of variability were similar. No significant differences in SOM contents were observed between Alfisols from the tropics vs. Alfisols of the temperate region, Ultisols from the tropics vs. Ultisols of the temperate region, and Mollisols of the tropics vs. Mollisols of the temperate region (Sanchez et al., 1982). Furthermore, no significant differences in organic matter contents were found between the classic black Mollisols or Chernozems of the temperate region and the red, highly weathered Oxisols of the tropics (Table 2). The high variability normally associated with SOM content (Wilding and Dyers, 1979) make differences difficult to detect at broad groups of soils.

Soil organic matter content is a function of additions and decomposition rates. Calculations from Greenland and Nye (1959) suggest that generally higher SOM decomposition rates found in the humid tropics caused by high temperatures and ample moisture are balanced by higher litter input, both factors being about five times higher in soils from tropical forests than from temperate forests (Sanchez, 1976). Many processes operating at site-specific rates affect actual input and decomposition rates and provide a wide range of equilibrium SOM contents (Anderson and Swift, 1983). It is safe to assume

Table 2. Mean total carbon and nitrogen reserves of soil orders in the tropical and temperate regions (Source: Sanchez et al., 1982.)

Region	Soil order	No. of profiles samples	Total C (kg C m^{-2})	
			0-15 cm depth	0-100 cm depth
Tropics	Oxisols	19	3.8 a	11.3 a
	Alfisols	13	2.9 a	6.4 b
	Ultisols	18	2.1 b	6.4 b
Temperate region	Mollisols	21	3.3 a	10.1 a
	Alfisols	16	2.8 ab	5.8 b
	Ultisols	8	2.4 b	4.2 b

therefore that the range in SOM contents in the tropics is as variable as in the temperate region. A more extensive study by Post et al. (1985) with over 3000 profiles arrived at the same conclusion. Unfortunately, there is no direct correlation between SOM contents and soil fertility as measured by plant productivity, other factors being constant. Mollisols shown in Table 2 supported many crops of corn without fertilization for years in midwestern U.S., Europe, and Argentina (Stevenson, 1984). But many Oxisols with similar SOM contents will definitely not be able to keep one corn plant alive without fertilizer additions in the Cerrado of Brazil (Lopes, 1983). Although this discrepancy is related to deficiencies of nutrients other than N and Al toxicity in Oxisols, it is clear that similar SOM contents do not guarantee adequate soil fertility. The processes governing the relationships between SOM and plant productivity need deeper understanding. One promising approach is to look at SOM fractions or functional pools (Sanchez and Miller, 1986). We believe that soil temperature, soil moisture regime, texture, and clay mineralogy are the main determinants of soil organic matter in the tropics.

The role of soil fauna is probably one of the most intriguing aspects of SOM in the tropics, which is in much need for quantification. Biopedoturbation is reported to operate through a much greater depth in humid and subhumid tropical soils than in temperate soils (Lal, 1987). The vector for horizon mixing is termites (Young, 1976; Russell, 1981), though the role of tropical earthworms and *Atta* ants is just beginning to be understood (Lavelle, 1984). The net result of horizon mixing by these organisms is not well understood. However, large quantities of subsoil are regularly brought to the soil surface by termites, and in the Neotropics, *Atta* ants incorporate considerable amounts of fresh leaf material into mineral soil at depths sometimes exceeding 1 meter (Haines, 1978).

Dynamics of SOM in Tropical Ecosystems

The CENTURY SOM model has been used to simulate SOM dynamics for a large number of different grassland sites in the United States (Parton et al., 1987) and a variety of different agroecosystems in the U.S., Canada, and Sweden (Parton et al., 1983; Parton et al., 1988; Cole et al., 1989). Our extensive experiences with testing and adapting the CENTURY model at a large number of sites suggests that this would be an excellent model for use in the tropics. To test the generality of the model for tropical sites, we selected a tropical forest site in Hawaii (Vitousek et al., in press), a tropical grassland site in Africa (McNaughton, 1983), and an agroecosystem site in Peru (Sanchez et al., 1983). These sites were selected because of the existence of adequate soils, and plant production data bases. We will describe the process used to adapt the CENTURY model for these sites and will present model results from these sites. The results will demonstrate usefulness of the model for simulating SOM dynamics of tropical sites, suggest needed improvements to the model, and hopefully help direct new research efforts on the management of tropical soils.

Model Description

The CENTURY SOM model was developed to simulate the dynamics of SOM for natural grassland systems and agroecosystems developed from grassland systems. The model simulates the cycling of carbon (C), nitrogen (N), phosphorus (P), and sulfur (S) in the plant-soil system. One of the key features of the model is that SOM is divided into three different fractions which have different turnover times (0.14, 5, and 150 years, respectively, under ideal conditions). This type of model structure is similar to that used for most of the recently developed SOM models (Jenkinson and Rayner, 1978; Paul and van Veen, 1978; van Veen and Paul, 1981). Three SOM fractions which we have identified include: (1) an active soil fraction consisting of live microbes and microbial products (0.14-year turnover time under ideal conditions); (2) a protected fraction that is more resistent to decomposition (5-year turnover time under ideal conditions) as a result of physical or chemical protection; and (3) a fraction that is physically or chemically resistant with a long turnover time (150-year turnover time under ideal conditions). The theoretical basis for these fractions is discussed in a paper presented by Anderson (1979).

The major processes represented by the model include plant production, mineralization of N, P, and S cycling of both organic and inorganic N, P, and S soil compounds, decomposition, water flow, and plant nutrient uptake. The flow diagrams for the C, N, and P submodels (Figures 1, 2, and 3) show that the structure of the submodels is similar with the exception that the P submodel includes inorganic P variables and flows not needed for the C and N submodels. The S submodel was not used for this set of model runs. Incoming plant residue from roots and shoots is divided into structural and metabolic pools as a function of the lignin to N ratio of the residue. Decomposition of the

Figure 1. Flow diagram for the carbon submodel. (Adapted from Parton et al., 1987.)

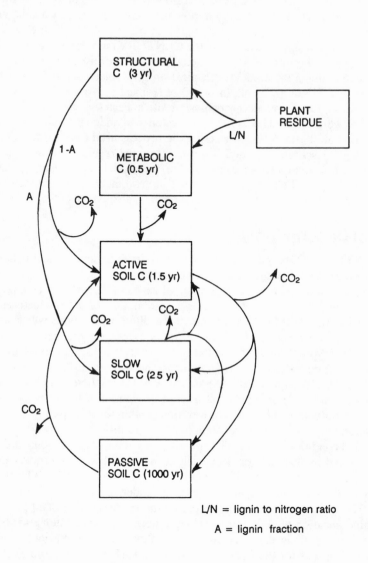

L/N = lignin to nitrogen ratio

A = lignin fraction

Dynamics of SOM in Tropical Ecosystems

Figure 2. Flow diagram for the nitrogen submodel. (Parton et al., 1987.)

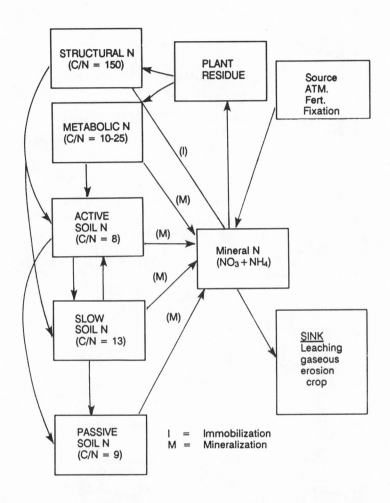

state variables is calculated by multiplying the decay rate specified for each variable times the combined effect of soil moisture and temperature on decomposition (calculated as a function of monthly maximum and minimum air temperature and precipitation). Decomposition rates are increased during months with soil tillage. The decay rates of structural plant residue are also a function of the lignin content of the structural pool (lower for high lignin content), and active SOM decay rates change as function of the soil texture (higher for sandy soils). There is a respiration loss associated with each of the

Figure 3. Flow diagram for the phosphorus submodel. (Adapted from Parton et al., 1988.)

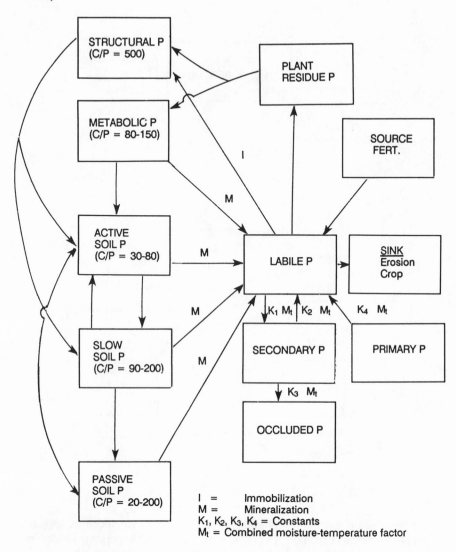

carbon flows which is fixed for all of the flows except for active SOM which varies with soil texture (increasing for sandy soils). Monthly plant production is estimated by assuming that maximum plant production is a function of annual precipitation and reduced if sufficient mineral N or P is not available. The C/N and C/P ratios of newly produced plant material are allowed to float within the specified range. Plant production is reduced if available N or P are insufficient

Dynamics of SOM in Tropical Ecosystems

to produce plant material with C/N or C/P ratios less than or equal to the maximum values of these ratios. Production will be limited by the element that is most limiting. The specified C/N and C/P ratios are input variables for the model and are plant species specific.

The nutrient submodels have the same basic structure as the carbon flow diagram and we calculate the nutrient flow associated to the flow of C by multiplying the C flow rate by the appropriate N/C or P/C ratio. Given the N/C and P/C ratios of the state variables and the CO_2 flow associated with each C flow, you find that net nutrient mineralization occurs as a result of decomposition of active, slow, and passive SOM, while decomposition of structural plant material (high C/N and C/P ratios) results in nutrient immobilization. The C/N ratios of the slow and passive SOM fractions are fixed in the model while the C/P ratios float as a function of the available P in the soil (low ratios for high P levels). The C/N and C/P ratios are input variables for the model which were modified from the values shown in Figures 2 and 3 in accordance with the observed values.

The dynamics of the mineral forms of N and P are represented by the model. In the N model, NO_3 leaching, atmospheric N inputs, soil N fixation, and the addition of N fertilizer are all represented. The P model includes the weathering of parent P, flows between secondary and labile P, formation of occluded P, and the addition of P via fertilization. In the present form of the model, soil texture and pH have an effect on the equilibrium between secondary and labile P, and soil texture modifies the weathering rate of primary P. We anticipate that more advanced versions of tropical SOM models will have to make further modifications to the mineral cycling part of the P submodel. We anticipate that soil mineralogy will impact on P fertilizer fixation by clay minerals, the solubility of secondary P, and the rate of formation of occluded P and that the impact of pH on cycling of mineral P will have to be modified. The complete description of the C and N part of the model is presented by Parton et al. (1987), while the P and S submodels are documented by Parton et al. (1988) and Parton et al. (1989).

The major input variables required for the CENTURY model include: (1) the monthly maximum and minimum air temperature; (2) monthly precipitation; (3) soil texture (sand, silt, and clay content); (4) the C/N and C/P ratios of dead shoot and root material; (5) lignin content of shoots and roots; (6) estimate of the atmospheric N inputs and soil N fixation rates; (7) estimated maximum plant production without nutrient limitations; and (8) the initial estimates of C, N, and P for inorganic and organic soil pools. If the monthly temperature and precipitation data and soil texture are known, then all of the other input variables can be estimated by the model for temperate grasslands and agroecosystems (equilibrium model runs and generalized relationships can be used).

Model Simulations

We selected three tropical sites to test the ability of CENTURY to simulate SOM dynamics for tropical soils. The sites include three tropical montane forest sites in Hawaii that developed on volcanic lava material during the last 3200 years, and an agroecosystem site near Yurimaguas, Peru, and two different sites within the Serengeti grasslands in Tanzania. These sites were selected because of the existence of adequate soils and plant production data to parameterize and test the model's ability to represent different types of natural and managed systems. In this section, we will discuss the model parameterization procedure and present simulated model results for appropriate model runs. We will attempt to show how well the model works for these sites and will be focusing on the potential modifications needed to simulate the dynamics of tropical SOM.

Hawaii Forest Sites

We selected three paired plot sites (Vitousek et al., in press) along elevational and age gradients in Hawaiian montane rainforest to test the ability of the model to simulate successional changes in SOM levels and plant production. The sites are located on an elevation gradient which ranges from 760 to 1675 m on the east slope of Mauna Loa volcano upslope from the city of Hilo, Hawaii. One set of sites is located on an 1855 lava flow while the other set of sites is located on an older Punahoa flow (^{14}C dates at 3110 years before present). Along the elevational gradient the mean annual temperature decreases from 18°C at the lower site to 12°C for the high site, while the annual precipitation decreases for 6350 mm at the lower site to 2700 mm at the high site. At each of the two aged lava flows and three elevations, Vitousek et al. (in press) have collected soil C and nutrient content data and have also sampled foliar nutrient content for the dominant plant species (*Metrosideros polymorpha*) that is found on all the sites. All the sites are classified as Histosols.

The CENTURY was set up to simulate a 3200-year sequence of plant production and SOM development for the three sites (elevations of 760, 1220, and 1675 m). The climatic data was estimated from the Atlas of Hawaii (1983). The C/N and C/P ratios for plant material and soil organic matter were estimated by using the observed values from the sites (Vitousek et al., in press); the root to shoot ratio for plant production was assumed to be 0.75, and the lignin content of the vegetation was assumed to be 15%. The N inputs (0.8, 0.68, 0.56 gm^2, respectively for the low, medium, and high elevation sites) were estimated by Peter Vitousek (personal communication) and weathering of P from the lava parent material was based on the observed accumulation of P in the soil during the first 135 years of soil genesis. We assumed that the formation of passive SOM would be minimal for soils that weather from the lava parent material. Preliminary model results for the Hawaii sites showed that SOM formation would be greatly overestimated if we assumed that passive SOM is formed in these soils. We also changed the equation which we used to represent the impact of labile P on the C/P ratio of slow SOM. This change was necessitated by the very high C/P ratios observed at these sites.

Hawaii Model Results

The results show (Figures 4a, b, c) that increases in soil C and N with time follows the same pattern as the plant production curve. We assumed that most of the SOM is composed of fairly active fractions (< 30-yr turnover time) and thus accumulation of SOM is expected to reflect changes in carbon input rates to the soil. A comparison of the observed and simulated soil C, N, and P levels (Table 3) after 135 years of soil formation (young flow) show that soil C, N, and P levels decrease with increasing elevation. The model suggests that this pattern results from higher P availability (higher weathering rates) with decreasing elevation, which leads to higher plant production. After 1000 to 2000 years of soil formation the pattern is reversed with the low elevation site having the lowest soil C and N levels. The higher soil C and N levels at the higher elevation sites results from abiotic decomposition rates which decrease from 0.48 kg yr^{-1} at the low elevation site to 0.26 kg yr^{-1} at the high elevation site. The lower decomposition rates result in higher soil C and N stabilization in spite of lower plant production at the higher sites. The only major discrepancy between the model results and observed data is that the model overestimates the soil C and N levels for the high elevation site after 135 years of soil formation. This discrepancy is due to an overestimate of P weathering rates for the high elevation site, which results in simulated plant production levels that are too high for the high elevation site. Preliminary plant production data from the site (Vitousek, personal communication) suggest that the simulated overall pattern of decreasing production with increasing elevation is consistent with the limited observed data.

A comparison of the observed and simulated soil P levels shows that observed soil P levels are highest at the intermediate site followed by the high and low elevation sites, while the model results show that soil P decreases with increasing elevation. The model results occur because of the assumptions that soil P weathering rates are proportional to abiotic decomposition rates (decreasing with elevation because of lower temperatures), and that P losses from the system are minimal. A possible explanation for the observed low total P levels at the low elevation site is that high annual rainfall (twice as great as the high elevation site) leaches organic P compounds from the system. J. W. B. Stewart (personal communication) suggests that organic P compounds are more mobile than inorganic P compounds in the soil and that the intense and frequent rainfall amounts could result in the flushing of organic P compounds through the soil. Being Histosols, it is assumed that most of the soil P will be organic P.

Figures 4a, b, c. Simulated time series of soil development for 3000 years of soil formation for low, intermediate, and high elevation sites. Aboveground production total soil C, and total soil P are represented for this time series.

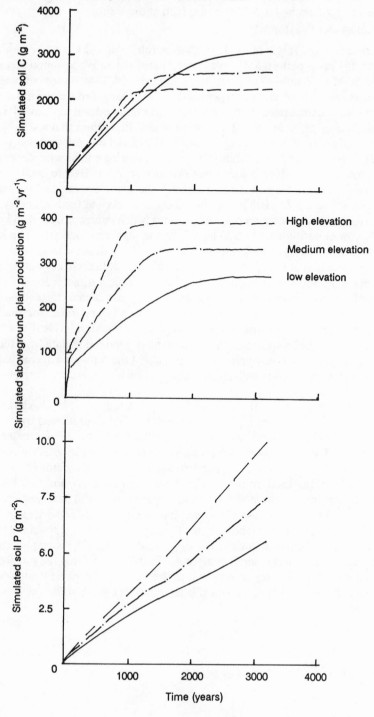

Table 3. Comparison of observed (± standard errors in parentheses) and CENTURY-simulated changes on soil C, N, and P content (g²) after 135 and 3100 years of soil formation at low (750 m), medium (1220 m), and high elevation (1675 m) lava flow sites. Soils are Histosols of the Island of Hawaii.

Site	Elevation	Soil C (g²)		Soil N (g²)		Soil P (g²)		Aboveground plant production (g² yr⁻¹)	
		Observed	Simulated	Observed	Simulated	Observed	Simulated	Observed	Simulated
Young flow (135 yr)	Low	1220 (180)	686	34 (5)	27	1.0 (0.2)	1.00	—	146
	Medium	900 (320)	661	21 (8)	26	0.7 (0.3)	0.95	—	111
	High	49 (27)	634	3 (2)	—	0.1 (0.03)	0.89	—	78
Old flow (3100 yr)	Low	1830 (210)	2242	53 (4)	77	3.1 (0.4)	10.2	400–600	390
	Medium	2890 (330)	2621	113 (12)	90	9.2 (1.3)	7.6	—	332
	High	2850 (430)	3071	79 (16)	105	6.0 (1.1)	5.8	—	269

Tropical Agroecosystem Site

The tropical agroecosystem site we selected is the Yurimaguas Experiment Station, Peru. The average annual rainfall is 2100 mm with little pronounced seasonal variation. Seasonal variations in air temperature are minimal and the average maximum and minimum daily air temperatures are 31 and 21° C, respectively. The Yurimaguas loam Ultisol has a low capacity to retain nutrients, low pH (4.5), aluminum toxicity, and low K reserves. A 17-year old secondary forest was slashed, burned and planted using a rice, corn, and soybean rotation with or without fertilization for 16 years (Sanchez et al., 1983; Sanchez, unpublished data). The check plot did not receive any fertilizer or lime, while the fertilizer treatment included complete fertilization (N, P, K, Ca, Mg, S, Zn, Cu, B, and Mo) and lime.

The CENTURY model was used to simulate the check and fertilizer (160 kg yr^{-1}) treatments. We simulated plant production, C and N SOM dynamics, and were unable to simulate P SOM dynamics because of the lack of an adequate data base to initialize the P state variables. The input variables for the model included the long-term average monthly maximum and minimum air temperature and precipitation, initial soil C and N levels, and the soil texture. Some of the other parameters which were initialized include the maximum plant production, C/N ratio, lignin content of the plant material, and estimated N inputs from the atmosphere (16.7 kg N yr^{-1}).

Tropical Agroecosystem Model Results

Both the observed data and simulated results (Figures 5a, b) show that soil C levels (0 to 20 cm) decreased with time after cultivation started. As expected, soil C losses were greatest for the control run. A comparison of the model results (Figure 4a) with the observed data (1 standard deviation) suggest that the model did a reasonable job of simulating the observed changes in soil C. Both observed and simulated results followed a pattern similar to the soil N levels (data not shown). Simulated grain yield for the fertilizer run was 6 t ha^{-1} during the 20-year run, while yields for the control run were initially 4.5 t ha^{-1} and then decreased to 2 t ha^{-1} after 10 years and 1 t ha^{-1} after 20 years. The observed data for the fertilizer case (6 to 7 t ha^{-1} yr^{-1} Sanchez et al., 1983) compares well with the simulated results. However, the model overestimates the production for control runs by 1 t ha^{-1} yr^{-1} from year 3 to 16 after cultivation was started. These results suggest that the observed decreases in plant production with time for the control plots is substantially underestimated by the model. A possible reason for these results is that the model does not consider the effect of other limiting elements (P, K) on production nor did we consider the direct impact of low soil pH (4.5) and Al toxicity of the control soils on plant production. These are factors which need to be considered in more refined tropical SOM models.

Figure 5a, b. Simulated plant production and soil C levels for a 20-year simulation of fertilizer and control runs at an agroecosystem site in Yurimaguas, Peru. The observed soil C levels ($\pm$ 1 standard deviation) are also presented.

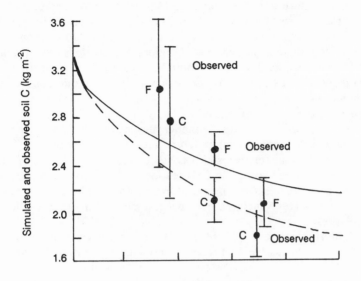

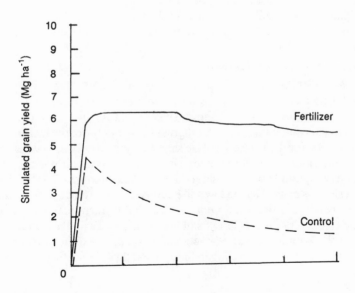

Tropical Grasslands Sites

We set up the model to simulate two tropical grassland sites within the Serengeti National Park, Tanzania. One site was in the shortgrass plains region which is heavily grazed during the November to May rainy season and lightly grazed during the dry season (June to October). The other site is located in the tallgrass savanna region where the grazing pressure is lower and fire is frequent. The annual precipitation is greater (900 vs 600 mm) at the tallgrass savanna site where the rainy season lasts from November to July. The soil texture for the shortgrass plains site (Calcomstole Setrcalciustoll) is loamy and soil P levels are quite high, while the tallgrass savanna site (Udic Hoplustoll) has a sandy loam soil texture and very low soil P levels.

The data used to set up the model runs include the long term monthly maximum and minimum air temperature and precipitation data, soil texture data, total soil organic matter, soil N content and soil P content data, and the lignin content of the aboveground vegetation, and estimates of plant production. The above data set was provided by S. J. McNaughton and is based on recent data sets (McNaughton, personal communication) from his Serengeti research sites.

Unfortunately, estimates of N inputs to the system were not available for the region. The values we used were substantially higher than the values used for U.S. Great Plains soils (Parton et al., 1987) because higher values were needed to get the model to correctly simulate the observed production at the shortgrass plains site. The model was set up to simulate the steady state plant production and SOM levels for the two sites by running the model for 2000 years. The observed soil C, N, and P levels were used to initialize the SOM state variables. None of the internal parameters were changed for these simulations.

Tropical Grasslands Model Results

A comparison of the observed and simulated soil C levels (Table 4) shows that the model correctly simulated the observed patterns of lower soil C for the tallgrass savanna site. Data not shown suggest that the high sand content and higher decomposition rates for the tallgrass savanna site are the factors causing the lower soil C levels. The results also show that the model overestimated plant production for the tallgrass savanna site. These results do not support our original assumption that the observed lower plant production at the tallgrass savanna site was caused by the low soil P levels. Some problems with interpreting these results are that we do not have the soil P fractionation data needed for good estimates of the different soil P variables (organic P, secondary P, and occluded P). Without having P fractionation data, it is difficult to know how much of total P is actively cycling in the soil and thus supporting plant production.

Table 4. Comparison of observed (+ standard errors in parentheses and CENTURY-simulated plant production and soil C levels for sites in Serengeti National Park, Tanzania. Soils are Ustols.

Site	Aboveground plant production $(g^2 yr^{-1})$		Soil C (g^2)	
	Observed	Simulated	Observed	Simulated
Shortgrass plains	497 (+187)	560	5558-9600	4328
Tallgrass savanna	400 (+93)	875	2700-5522	3000

Another discrepancy between the observed data and model results is the underestimate of soil C levels for the shortgrass plains site. The reason for this discrepancy is unclear. However, experience with using the CENTURY model shows that simulated soil C levels are sensitive to the efficiency of stabilizing active SOM (microbes and microbial products) into slow SOM and passive SOM rates. The efficiency of stabilizing active SOM into slow SOM has been shown to be a function of soil texture (Sorenson, 1981) and is likely to be a function of soil mineralogy which is different from soil mineralogy of soil used to parameterize the CENTURY model. The rate of formation of passive SOM is a process about which we have little information and in general have to depend on information which results from radioactive carbon dating of SOM fractions (i.e., Anderson and Paul, 1984). Unfortunately, radioactive carbon dating has only been done for a limited number of soils and much more data are needed for tropical soils.

Summary

The results show that existing SOM models need to be upgraded in order to simulate SOM dynamics, nutrient cycling, and plant production in tropical agroecosystems, forests, and grasslands. The results suggest that many of the important soil processes in the tropics are not quantified and that data needed to initiate the model input variables are hard to find. The preliminary model runs and our general knowledge of tropical soil management suggest that there are four major areas of research which would greatly advance our ability to simulate SOM dynamics for tropical soil. These include: (1) improving our understanding about the effect of soil texture, clay mineralogy, and iron and aluminum chemistry on SOM formation and decomposition; (2) advancing our knowledge about organic P dynamics in tropical soils; (3) studying the effect of clay mineralogy and parent material on the formation of passive SOM fractions;

and (4) evaluating the direct effect of low soil pH and aluminum toxicity on microbial processes such as nitrogen fixation and microbial decomposition.

Conclusions and Recommendations

Theme Imperative: An emphasis must be placed on increasing our ability to simulate SOM dynamics in the tropics through accumulation of an adequate database.

Research Imperatives:

1. Simulation models such as the CENTURY model require identification of three pools and their associated turnover rates. Current chemical characterization of SOM does not identify pools useful for modeling. Methods of characterizing active, slow, and passive pools and transfer rates are needed for further applications of the CENTURY model in both tropical and temperate soils. Further measurements of C, N, S, and P pools and transfer rates are needed in a variety of soil mineralogies and under several moisture and temperature regimes.

2. Soil fauna need to be explicitly included in models such as CENTURY. At present, fauna are implicit; activities can be accounted for by changes in values for parameters. Effects of soil fauna are likely to be most important in affecting turnover rates in slow and passive pools, in contrast to the active pool of carbon. Acceptable generalizations by experts on tropical soil fauna must be made explicit to modelers.

3. The highly aggregated character of many tropical soils and its consequences on porosity and pore-size distribution needs to be explicitly considered in modeling. Because of the highly aggregated character of many tropical soils, traditional analysis of particle-size distribution is misleading and not generally applicable. Experimental approaches need to be developed to measure physical and structural characteristics. The effects of aggregation on decomposition and turnover rates must be considered. The diverse mineralogy of tropical soils and its effect on decomposition and turnover should also be explicitly considered (e.g., the differences between kaolinitic, oxidic, and allophanic mineralogies). Further information on the extent and mechanism of soil organic matter influence on soil structure and aggregation is needed.

4. A critical parameter in the CENTURY model is the turnover time for the slow and passive pools. Existing estimates of rates are based on long-term incubation studies (5 years where the first year is excluded). New long-term studies are needed, based on realistic tropical conditions perhaps using tracer methods, including natural abundances.

5. Temperature and moisture relationships with metabolic rates may differ quantitatively between temperate and tropical soils, since biota are

adapted to different climatic regimes. Submodels of metabolic rates under tropical conditions should be developed to contribute to CENTURY model estimates of C, N, and P dynamics. The rates and pathways of the decomposition of organic matter undoubtedly differ among many tropical and temperate ecosystems. While the biota of soils are necessarily adapted to their particular climatic regime, the response of the microflora to extremes in their environment are very likely different.

6. The present model focuses primarily on C, N, and P dynamics. Nitrogen dynamics is implicitly modeled by the direct level with C cycling through the application of set C/N ratios. To complete the model to copy with C, N, P, and S dynamics, specific nitrogen cycling processes such as nitrification should be included.

7. Additional soil parameters, such as pH and soil structure, might profitably be included in the CENTURY model. The effects of soil nutrient status on decomposition rates must be examined experimentally. Because tropical soils encompass a variety of pH values, information on decomposition rates under different pH regimes is needed to decide whether direct effects of pH should be included in the CENTURY model.

8. In some tropical soils, organic matter becomes distributed to depths of several meters or more. Modification of the CENTURY model may be needed to allow for storage and turnover in deeper tropical soils and might include transport of organic matter by fauna.

9. In new modeling efforts, further attention should be paid to soil solution chemistry: leaching and exchange of soil nutrients are mediated by solution. Soil organic matter is important in cation exchange capacity, a parameter that might be predictable by further development in the CENTURY model. Information on soil CO_2 levels might be derived from existing models when information on soil atmosphere diffusion becomes incorporated in the model.

10. We recommend that a modeling workshop be held to demonstrate and improve the use of CENTURY for tropical systems. In the workshop, we would invite participation from groups that have extensive data sets for tropical agroecosystem and native forest sites. The CENTURY model would be set up to simulate sites before the workshop and workshop participants would learn how to use CENTURY and evaluate the usefulness of the model. We would set up the model so that we could make changes to the model in accordance with the results of the workshop.

Acknowledgments. We would like to thank Peter Vitousek and Sam McNaughton for providing data from their sites and evaluating the scientific merit of the nodule simulations.

Chapter 7

Methodologies for Assessing the Quantity and Quality of Soil Organic Matter

Frank J. Stevenson and Edward T. Elliott

with C. Vernon Cole, John Ingram, J. Malcolm Oades,
Caroline Preston, and Phillip J. Sollins

Abstract
Various fractions or pools of organic matter (OM) considered important in carbon (C) and nutrient cycling are described with comments on the utility of methods useful for measuring these pools. Emphasis is given to those fractions or pools that have biological significance. The turnover of soil organic matter (SOM) depends upon both kind of OM and its location in the soil, and any meaningful fractionation scheme must take into account both physical location and chemical composition. In continuously cropped soils, stabilized OM (humus) functions more as a "reservoir" of N, P, and S than as a "source" of these nutrients, but considerable interchanges occur with available pools and these interchanges are of importance in nutrient cycling. Two relatively new approaches for the *in situ* examination of SOM (^{13}C-NMR, and pyrolysis/mass spectrometry) are discussed.

Organic matter plays a major role in the productivity of soils and is particularly important as a reservoir of nutrients for biota in strongly weathered soils of the tropics. Organic matter: (1) is the energy source for heterotrophic life in soils, (2) is a major reservoir and source of N, P, S, and other nutrients required by plants, (3) has a major influence on soil pH, cation exchange capacity (CEC), and anion exchange capacity (AEC), and (4) plays a major role in soil structure and therefore various soil physical properties.

Soil organic matter is in a dynamic state. Inputs to soil occur from above as leaf litter, woody litter, epiphytes, insect and animal debris, and dissolved organic C from canopy drip. Assimilated C is moved belowground in root systems, including exudates, and by soil fauna. Some organic materials are utilized and "turned over" quickly, but some remain in the soil for long periods of time. Turnover is affected by interactions involving inorganic soil components, as well as by location of OM in zones (i.e., micropores) inaccessible to microorganisms and their enzymes.

Annual additions of C to most soils are matched by losses due to respiration and leaching and a steady state exists so that the content of OM remains constant. However, this steady state is disturbed by various management practices and it is important to realize that "turnover times" of OM in tropical soils are much shorter than in temperate soils. Thus, losses of OM will occur more quickly when tropical systems are disturbed than would be predicted from experiences in temperate regions.

There is an urgent need for realistic methods to assess the quantity and quality of SOM. Some fractionation of the complex and dynamic mixture of organic material in soils is required. Hopefully, the fractions will have some relation to the major biological entities in soils. This review outlines methods currently available to characterize SOM. Inevitably, the techniques are dominated by chemical characterization, but recently more biological and/or physical approaches have been applied.

The terms *humus* and *SOM* are often used synonymously. In this chapter, SOM will be used in a broad sense to include the whole of the OM in soils, including litter, light fraction, biomass, and soluble organics.

Chemical Fractionation

Much of the OM or humus in soils consists of a series of acidic, yellow to black macromolecules referred to as humic substances. The remainder consists of various classes of biochemical compounds, such as carbohydrates, lipids, and proteins which are referred to collectively as nonhumic substances. Humic and nonhumic substances are closely associated and are not readily separated from each other.

Humic substances represent a complex mixture of molecules having various sizes and shapes, but no completely satisfactory scheme has been

Dynamics of SOM in Tropical Ecosystems

forthcoming for their isolation, purification, and fractionation (Stevenson, 1982a; Hayes, 1985; Swift, 1985b; Parsons, 1988).

Extraction of SOM is commonly done with caustic alkali (usually 0.1 to 0.5 N NaOH), although, in recent years, use has been made of mild reagents, such as neutral $Na_4P_2O_7$. Repeated extraction is required to obtain maximum recovery of OM. The following fractions are commonly defined: *humic acid (HA)*, soluble in alkali, insoluble in acid; *fulvic acid (FA)*, soluble in both alkali and acid; and *humin*, insoluble in both alkali and acid.

Identification of the acid-soluble material as "FA" is questionable because this fraction invariably contains organic substances belonging to the well-known classes of organic compounds. The term FA should be reserved as a generic name for the pigmented components of the acid-soluble fraction, in which case the soluble material remaining after removal of HA should be referred to as the *FA fraction* (Stevenson, 1982a).

The solubility of humic substances in alkali is believed to be caused by disruption of bonds binding the molecules to clay and polyvalent cations; salts of monovalent cations (Na^+ and K^+) are soluble, whereas those of di- and trivalent cations are not. Leaching the soil with dilute HCl, which removes Ca^{+2} and other polyvalent cations, increases the efficiency of extraction. However, a certain amount of OM, normally less than 5% of the total for surface soils but somewhat more for subsoils, is extracted during this pretreatment.

Humic acids extracted with NaOH (or other reagents) nearly always contain a considerable amount of inorganic material (about 25%), much of which can be removed by repeated dissolution and precipitation, by passage through an ion-exchange resin column, or by use of HF to dissolve silicates. Substantial C is lost during these procedures.

Crude HA preparations may also contain variable amounts of coprecipitated or coadsorbed nonhumic substances (e.g., carbohydrates and proteins). Because these constituents are not easily removed by mild separation methods, it is sometimes thought that they are covalently bound to and are an integral part of HA. Methods for removing carbohydrate and protein constituents include hydrolysis with dilute mineral acids, enzymatic hydrolysis, gel filtration, and phenol extraction. In the latter case, treatment of HA with aqueous phenol has led to the separation of a protein-rich component (McGill and Paul, 1976).

Similarly, a range of "purification" procedures can be applied to the pigmented components of the acid soluble fraction. A popular approach for recovery of FA's from natural waters is by sorption-desorption from a macroreticular resin (i.e., XAD-8), and this method shows promise for recovery of generic FA's from soils (Aiken, 1988). A fractionation scheme leading to recovery of generic HA's and FA's free of inorganic and organic contaminants is given in Figure 1.

Figure 1. Extraction and fractionation of humic and fulvic acids. (Adapted from Kumada, 1987, page 73.)

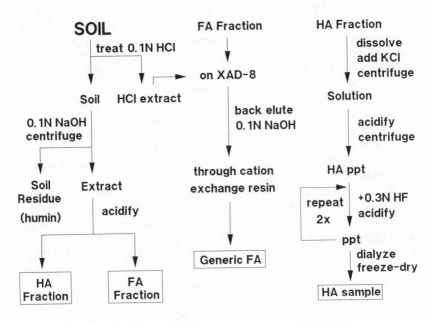

The humin fraction may consist of one or more of the following: (1) HA's so intimately bound to mineral matter that the two cannot be separated, (2) highly condensed (humified) humic matter with a high C content (> 60%) and thereby insoluble in alkali, (3) fungal melanins, and (4) paraffinic structures (Hatcher et al., 1985). It should be noted that undecomposed plant material, if not removed before alkali extraction, will also contribute to the humin fraction. Also, some contribution can be made by charcoal left behind through burning of plant residues (such as in sugarcane fields). Concentration and enrichment of humin can be done by repeated acid and alkali extraction followed by treatment of the final soil residue with an acid solution containing HF.

Genetic Relationships of Classical Fractions

Genetic relationships between the various humic components (HA's, FA's, and humins) remain unknown and to some extent unexplored. One hypothesis is that the various fractions represent a heterogeneous mixture of molecules (system of polymers) that are formed by a common biochemical pathway. In any given soil, they range in molecular weight from as low as several hundred to perhaps over 300,000 Daltons. The postulated relationships are depicted in Figure 2, where C and O contents, acidity, and degree of polymerization, all change systematically with increasing molecular weight. The low-molecular-

Dynamics of SOM in Tropical Ecosystems

Figure 2. Classification and chemical properties of humic substances. Additional fractions of HA and FA are shown (see Stevenson, 1982a). Humin (not shown) is insoluble in alkali and consists of a complex mixture of substances (see text).

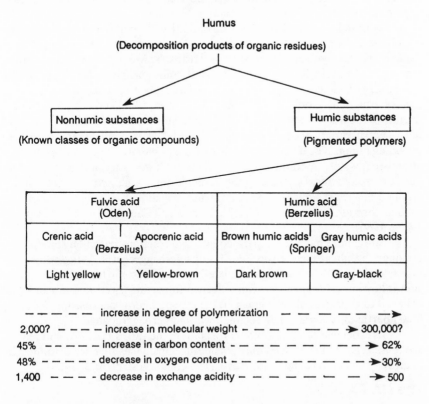

weight FA's have higher O contents but lower C contents than the high-molecular-weight HA's. Also, total acidities of FA's (usual range of 900 to 1,400 cmol (H^+) kg^{-1}) are considerably higher than for HA's (usual range of 500 to 870 cmol (H^+) kg^{-1}). Inasmuch as FA's are less susceptible to precipitation with acids and polyvalent cations, they are the constituents responsible for the brownish-yellow color of many natural waters.

A second hypothesis is that the various humic components represent substances that have been formed by different pathways, for which several are possible. The classical theory, popularized by Waksman (1936), is that humic substances represent modified lignins. Arguments in support of a degradative pathway for humus synthesis have been given by Hatcher and Spiker (1988). Many present-day soil scientists favor a polymerization pathway in which polyphenols (quinones) derived from lignin, together with those synthesized by microorganisms, polymerize alone or in the presence of amino compounds (amino acids, etc.) to form brown-colored polymers having variable molecular weights and contents of acidic functional groups. A third possible pathway is through condensation of amino acids and related substances with reducing sugars, according to the Maillard reaction.

In practice, all pathways may be operative, but not to the same extent in all soils or in the same order of importance. A lignin pathway, for example, may predominate in wet sediments, such as Histosols. Mention was made earlier that a portion of the humin, at least in some soils, may be derived from fungal melanins, as well as from partially decomposed plant remains. Results of ^{14}C studies using simple organic substrates (Stevenson, 1986) and of ^{15}N studies on the fate of recently immobilized N (Kelley and Stevenson, 1985; He et al., 1988a) have shown rapid incorporation of ^{14}C and ^{15}N into all humus fractions, including humin. The data obtained thus far suggest that each soil will ultimately attain steady-state conditions with regard to both amount and composition of the OM it contains. In terms of the geologic time scale, equilibrium will be attained in a short time (< 100 years) when the soil environment is modified, such as through a change in cultivation practice (Campbell, 1978; Stevenson, 1982a, 1986).

Soil humin has been separated into two fractions based on separations obtained by ultrasonic disruption of soil microaggregates (Almendros and Gonzalez-Vila, 1987). The humin isolated by the ultrasonic treatment (suspension following centrifugation) was referred to as "inherited humin" and was believed to consist of altered lignin-like polymers and/or microscopic (subcellular) particles of plant origin retained within microaggregates.

Irrespective of their mode of formation, the terms HA, FA, and humin are to be regarded as group concepts uniting substances that are dissimilar in many respects even though they may have a common structural pattern.

The HA/FA Ratio

Considerable emphasis has been given from time to time to the HA/FA ratio as a means of evaluating humus quality and the genetic relationships of the different humus forms to soil formation. This work is difficult to interpret because most published data are based on separations obtained by acidification of the initial alkaline extract rather than the purified preparations (outlined in Figure 1). Accordingly, HA/FA ratios recorded in the literature refer to the operationally defined *fractions* rather than to *generic* HA's and FA's and must be interpreted with this in mind.

The so-called HA/FA ratio of the surface layer of soils ranges from as low as 0.3 to over 2.5 (Kononova, 1966). Soil representative of the Mollisols have the highest ratios; Spodosols and Alfisols have the lowest. The HA/FA ratio usually, but not always, decreases with increasing depth. For reasons that are not clear, the OM of many black alkali soils consists almost entirely of humin.

Differences also exist in the general chemical properties of HA. The HA's of Spodosols and Alfisols appear to be less "aromatic" in nature, and to more closely resemble FA's, than do the HA's of Mollisols. Kumada (1987) has classified HA's into four types (A, B, Rp, and P) based on optical properties (light absorption, IR spectra). This work is discussed in Theng et al. (see Chapter 1).

Conclusions and Recommendations

1. Terminology continues to be a major problem in studies on SOM. Unfortunately, FA continues to be used as a generic term for the alkali-soluble, acid-soluble fraction when it is well-known that appreciable amounts of both low- and high-molecular-weight biochemical compounds are also present.

2. From the earliest studies of soil humus, an acidification step has been used to precipitate HA, both for purification and as a convenient means of recovering a significant portion of the extracted organic matter. The question that needs to be asked is this (Parsons, 1988): Has the precipitation step become so ingrained in methodology, and in the operationally derived definitions, that research progress on humic substances has been stifled? Perhaps other fractionation procedures could be devised that would provide more meaningful data on humus composition and thereby a better understanding of the role of humus in soil fertility. Alternate fractionation methods have been discussed by Swift (1985b).

3. A major goal in the study of SOM is to relate information regarding the kinds and amounts of the various components to soil productivity. The procedure described herein is rather arbitrary in that limited data are provided on pool sizes (i.e., labile vs. nonlabile components). The use of milder and more specific extractants is recommended. A sequence of extractions using reagents specific for defined biologically active component(s) may be preferred over more complete extraction with a single reagent, as is commonly done.

Physical Fractionation

Chemical fractionation and characterization methods have not proven particularly useful in following the dynamics of organic materials in soils, due in part to the complex nature of SOM but also because an enormous array of compounds exists in soil, ranging from recent plant material through a continuum of metabolic products of microorganisms to components of stable humus. In recent years, biologically significant fractions have been obtained by methods based on disaggregation of the soil before chemical characterization.

Physical fractionation procedures have been used in the following types of studies:

1. To recover the "light fraction," consisting largely of undecomposed plant residues and their partial decomposition products.
2. To establish the nature of organic matter in organo-mineral complexes.
3. To determine the types of organic matter involved in the stabilization of water-stable aggregates.

Physical fractionation of SOM can be discussed in terms of methods for soil disruption or dispersion and separation of the various size fractions or components. Many treatment combinations are possible, each yielding a unique result. Application of any given method or sequence of treatments can give results that will need to be interpreted differently, depending upon the nature of the soil. A review of mechanisms involved in the formation of stable soil aggregates has been given by Oades (1984).

Methods of Disruption

Sonication

Sonication without chemical treatment has been used extensively for soil disruption in preparation for particle size or density fractionation (Turchenek and Oades, 1979; D.W. Anderson et al., 1981; Tiessen and Stewart, 1983; Tiessen et al., 1983, 1984a, 1984c; Christensen and Sørensen, 1985, 1986; Balesdent et al., 1987, 1988; Catroux and Schnitzer, 1987; Christensen, 1987; Oades et al., 1987). Sonication produces heat and the container holding the sample should be encased in an ice bath. Even so, temperatures may reach 45°C at high energy levels and long sonication times (30 min). Three hundred watts for 15 min is usually sufficient for maximum dispersion. However, soils are highly variable and complete dispersion may occur for some soil types but not others (Churchman and Tate, 1986). For comparison of results, sonicators should be calibrated, which is difficult and time consuming and thereby seldom done (Christensen, 1985).

Increasing the intensity of sonication usually results in recovery of more OM in the fine fraction. Since all workers do not use the same or even known energy levels, results from different laboratories are not directly comparable.

Shaking

Shaking is a more gentle procedure for soil disruption. Disruption is more rapid and complete with sandy soils than with heavy-textured ones. Glass beads (Balesdent et al., 1987, 1988) or agate marbles have been used to increase the rate of disruption and to eliminate problems created by dispersion of a series of soils with variable sand contents. Sand content may also be a factor in the rate of disruption by sonication (Edwards and Bremner, 1967).

Chemical Pretreatments

Various chemical pretreatments have been used to aid dispersion or to remove a specific component responsible for the stabilization of soil aggregates.

Samples are often saturated with Na$^+$ by treatment with a Na-resin, pyrophosphate, or hexametaphosphate. Complexing agents have been used to remove polyvalent cations, dithionite to reduce and dissolve iron oxides, and periodate to selectively oxidize carbohydrates. The disaggregating agent usually results in the solubilization or oxidation of a select portion of the OM (Stefanson, 1971; Hamblin and Greenland, 1977; Cheshire et al., 1983, 1984, 1985; Churchman and Tate, 1986).

Separation of Fractions by Physical Means

Three basic methods of physical separation of soil have been used: sieving, sedimentation, and densitometry. Separation by sieving is based upon size (shapes may vary); separation by sedimentation is based on an equivalent spherical diameter of particles (which will vary in size, shape, and density); and separation by densitometry or weight per unit volume, which is independent of size or shape. These methods can be used in combination (e.g., sequential sieving and densitometry) to obtain, for example, particles with specific sizes and densities.

Sieving is usually used for particles $> 50 \mu$m in diameter; sedimentation is used under gravity for particles $> 2 \mu$m; and various centrifugation techniques are used for particles $< 2 \mu$m. Density separations can be applied to the whole soil, to particle-size fractions with density gradients, or to solutions with a specific density, usually from 1.6 to 2.2 Mg m^{-3}. Greatest concentrations of organic materials are usually obtained in the lightest fractions. The technique requires centrifugation to obtain isopycnic equilibria with considerable g forces involved when working with fine particles.

Sieving

Sieving can be used to separate various size fractions from soils. Both dry and wet techniques have been used. Dry sieving has often been used in wind erosion studies to determine particle sizes susceptible to wind erosion. The approach has also been used in some SOM studies (Christensen, 1986) and may be useful to separate organic debris from sands. However, dry sieving of aggregated soils is difficult to justify unless one is interested in the dry mechanical strength of aggregates.

Wet sieving is much more widely used with two main purposes. The first is to determine particle-size distribution of soils (mechanical analysis) after peroxidation of OM and addition of a dispersant such as Calgon (sodium hexametaphosphate) and NaOH to raise the pH to further enhance dispersion of fine particles. Various sand fractions are separated on sieves; silts and clays are separated by sedimentation. Sieving is used extensively to determine the stability of aggregates in soil. In the former case, a particular particle-size fraction is obtained by dry sieving and the stability of this size fraction is then determined by wet sieving. A particle-size distribution of water-stable aggregates is obtained by wet sieving the whole soil. In both case, the proportions of single-grain particles (e.g., sand) in the aggregate size fractions need to be determined. Standard methods for sieving are described by Kemper and Rosenau (1986).

Wet sieving is an empirical procedure but both manual and automatic systems can be standardized to give reproducible results. Water-stable aggregate stabilities then depend on the treatment given to the soil before sieving. This includes how the samples were taken, the water content at sampling, drying of the samples and subsequent wetting (rapid or slow), and any mechanical disturbances during prehandling procedures. Pre-treatments may include selective chemical treatments aimed to remove putative cements. Sieving can be used for separation of large aggregates from relatively undisturbed soil (Tisdall and Oades, 1982; Dormaar, 1983, 1984; Elliott, 1986). Excessive handling or processing prior to sieving will change the aggregate-size distribution of most soils. The use of intact soil cores is the least destructive method. It is recommended that, where possible, samples be passed through a large sieve (6-8 mm) before sieving to obtain more uniform results. Air drying will have drastic effects on aggregate-size distribution in many tropical soils, particularly in Andosols.

The disruption of larger aggregates (> 250 μm) by shaking is most severe when dry aggregates are wetted rapidly by direct immersion in water. The wet sieving of soils sampled in a wet state will yield greater values for aggregate stability than wet sieving of soils sampled air-dry. Some workers have attempted to maximize shaking by drying samples to some standard water content prior to wet sieving. This should not be done for tropical soils which often remain permanently wet. Other workers have attempted to bring soils to a standard water content before wet sieving by exposing the soil to a vapor mist or by wetting the soil under vacuum or controlled slow wetting on a pressure plate or controlled additions of water to absorbent paper supporting the soil (Hoffman and de Leenheer, 1975; Kemper and Rosenau, 1986). Care must be taken to prevent microbial activity during pretreatments that take several days.

For investigations of OM in aggregate fractions, it is advisable to recover all soil fractions for analysis to obtain a complete picture of OM dynamics (Tisdall and Oades, 1980; Elliott, 1986). The finest size fractions may be further fractionated by sedimentation.

Sedimentation

Sedimentation is based on Stoke's law, from which calculations are made for settling times for a given estimated spherical diameter (ESD) of a settling particle. Exceptions to Stoke's law include deviations from spherical shape and particle density used for the calculations. A particle density of 2.65 Mg m^{-3} is usually assumed for mineral particles, but it is clear that aggregates and organo-mineral associations have much lower densities. Using Stoke's law, it can be shown that a particle with a density of 1.4 Mg m^{-3} and a diameter of 0.04 mm would sediment at the same rate as a particle with a density of 2.65 Mg m^{-3} and a diameter 0.02 mm. In addition, there are shape factors that will influence the rate of sedimentation. Many organo-mineral particles are porous and when these pores are filled with water the density of the sedimenting particle is then the density of the solid plus water. Thus, fractions obtained by sedimentation are not homogeneous with respect to the type of organic matter they contain (Tiessen and Stewart, 1983; Balesdent et al., 1988).

For complete separation of any given size fraction, the process must be repeated several times. The fine material remaining in the supernatant can be recovered and further fractionated by centrifugation. Common fractions are sand, coarse silt, fine silt, coarse clay, and fine clay, although more complete separations are often made (Figure 3). Depending on the analytical measurements to be made, the various fractions are oven or freeze-dried. They can then be further separated densitometrically (Turchenek and Oades, 1979) or subjected to chemical, microbiological (Cameron and Posner, 1979; Ahmed and Oades, 1984), or microscopic examination (Tiessen and Stewart, 1988).

Density Separations

Separations based on density have been used for two main purposes in studies of SOM: to separate plant and animal debris in a "light fraction"

Figure 3. A typical scheme for the physical fractionation of soil organic matter. The texture classes are not those of international or U.S. standards. (Adapted from Turchenek and Oades, 1979.)

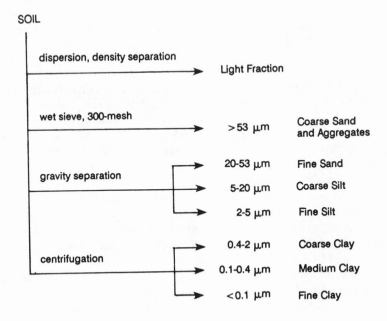

(Spycher et al., 1983; Sollins et al., 1984) and to separate organo-mineral associations, usually in aggregates (Turchenek and Oades, 1979).

Heavy liquids used for separation can be either inorganic (such as NaI, ZnBr, or CsCl) or organic (such as tetrabromomethane, tetrabromoethane, or dibromochloropropane). Organic liquids are potentially more dense, but they are dangerous to use because of toxicity. This is an important consideration if subsequent incubation of the isolated fractions is to be performed (Sollins et al., 1984), although heavy inorganic liquids can also affect microbial activity.

The use of heavy liquids involves a range of problems. For organo-mineral particle fractionations, the soil is generally dispersed, sedimented in water, dried, and suspended in an organic liquid that is miscible with water in order to further "dry" the particles. The liquid in which the particles are suspended may be miscible with the heavy liquid (i.e., acetone, Turchenek and Oades, 1979), in which case changes in density of the heavy liquid must be considered. Alternatively, the dried particles can be suspended in an organic liquid not miscible with the heavy liquid (i.e., ethanol, Spycher and Young, 1977), in which case the lighter liquid is removed from the surface of the heavy liquid after separation. Surfactants have been used to maintain suspension of fine particles that otherwise would coagulate in the heavy liquid (Turchenek and Oades, 1979). However, surfactants may adhere to soil particles, thereby resulting in analytical errors for organic C in the different fractions. Spycher and Young (1977) devised a method to avoid coagulation through repeated resuspension of the sample. Density fractions are removed by centrifugation in sequentially heavier liquids with material floating on the surface being removed each time as the light fraction. Particles may be resuspended several times at a particular density to obtain a more complete separation.

Organic Matter and Aggregation

Edwards and Bremner (1967) suggested that a large proportion of soil particles and OM is contained in microaggregates ($<250~\mu$m) consisting of clay and humified OM linked by polyvalent cations. Tisdall and Oades (1982) presented a conceptual model in which persistent OM binds primary minerals into small microaggregates (2-20 μm in diameter), (sensu Edwards and Bremner) that are stable to sonication. These in turn, are bound together by additional persistent OM (in association with polyvalent metals) to form larger size microaggregates (20-250 μm in diameter), which disintegrate with sonication but remain intact with long periods of cultivation. They further suggest that microaggregates are bound together to form macroaggregates (250-2000 μm), both by transient (polysaccharides) and temporary (roots and fungal hyphae) OM. Elliott (1986) and Elliott and Coleman (1988) suggested that this transient OM has a longer turnover time than that suggested by Tisdall and Oades (1982) and is that pool of OM which is lost due to cultivation. As a result, this OM pool is the primary contributor to the natural fertility of grassland soils. On the basis of a ^{13}C-NMR study of some cultivated Vertisols and adjoining virgin soils, Skjemstad et al. (1986) concluded that the major mechanism for the relative stability of persistent SOM was a physical association with inorganic components rather than as inherent chemical or biological inertness of the OM itself.

With increasing intensity of physical disruption, particle-size distribution patterns shift toward smaller particles (Tisdall and Oades, 1980; Dormaar, 1983; Elliott, 1986). Highest yield of large particles (aggregates) are obtained with wet sieving; highest yields of small particles are obtained with sonication. It follows that some particles obtained by sonication are associated with macroaggregates obtained by wet sieving, depending upon the soil and cultivation history.

The sand fraction obtained by sonication/sedimentation separation contains most of the light fraction OM, mainly undecomposed plant remains (Turchenek and Oades, 1979; Tiessen and Stewart, 1983; Christensen, 1987). The silt and coarse clay fraction is composed of microaggregates or parts of microaggregates plus primary particles of this class size (Oades and Turchenek, 1978); the fine clay fraction contain inter-microaggregate OM plus OM adsorbed to clay particles.

There is a large body of information supporting the quantitative importance and resistant nature of OM associated with the fine silt/coarse clay fraction. Organic matter of the fine clay, as obtained by sonication/sedimentation, has been shown to have a faster turnover rate than that of the silt and coarse clay fraction during decades of cultivation (Tiessen and Stewart, 1983; Dalal and Mayer, 1986a).

Carbon-14 dating of OM in the various particle-size classes has shown that OM of the coarse clay is the oldest and equivalent to the HA fraction of SOM; OM of the fine clay is somewhat younger (D. W. Anderson and Paul, 1984). The natural abundance of ^{15}N of the fine silt and coarse clay fraction was found by Tiessen et al. (1984a) to be relatively unchanged after 60 years of cultivation while other fractions varied in ^{15}N abundance. Based on studies of the natural abundance of ^{13}C, Balesdent et al. (1988) found that OM of the coarse clay was a heterogeneous pool with 50% turnover between 1888 and 1915, after which little change occurred for the next 71 years; this fraction showed behavior intermediate between the fine silt and fine clay.

After five years of incubation in the field, ^{15}N and ^{14}C from labeled substrates appeared more quickly in the clay fraction than in the silt fraction and all soils had higher proportions of labeled ^{15}N than native N in the clay, with the converse being true for silt (Christensen and Sørenson, 1986), indicating a faster turnover of organic N in the clay fraction. Longer term field studies with tracers indicate that "clay bound OM may be important in medium-term OM turnover, whereas silt-bound OM may participate in longer-term OM cycling" (Christensen and Sørenson, 1985).

Results of incubation studies to determine net mineralization rates of organic N, P, and S support the finding that differences exist in OM recalcitrance for the various size fractions. Less C and/or N become mineralized from the silt fraction than the clay fraction (Cameron and Posner, 1979; Lowe and Hinds, 1983; Christensen, 1987). Catroux and Schnitzer (1987) found that most of the mineralizable N was derived from the silt and coarse clay. Their soil may have been cultivated for many years and the fine clay fraction may have been depleted, as observed by Tiessen and Stewart (1983). However, Catroux and Schnitzer (1987) did find that the amount of N mineralized per unit of organic N was highest for the fine clay.

Based on these data, it is proposed that, for loams and clays, as cultivation intensity and time increases, inter-microaggregate OM is lost through mineralization and macroaggregates disintegrate into microaggregates. The net result is a shift from the condition in which macroaggregates are the dominant structural unit to that in which microaggregates are the dominant form, as observed by Tisdall and Oades (1980), Dormaar (1983), and Elliott (1986).

This model of the role of OM in aggregate hierarchy seem to be applicable to a number of soils in temperate regions, particularly those with strong wet-dry climatic regimes. It is clearly worth comparing the OM pools with the three conceptual pools used in the simulation model of Parton et al. (1987). These pools, with turnover times and suggested constituents, are: (1) active (1.5 years, microbial biomass and light fraction, roots and hyphae), (2) slow (25 years, transient, inter-microaggregate material mostly within macroaggregates), and (3) passive (1,000 years, intra-microaggregate material).

Is it possible to fractionate soils physically to obtain real measures of these pools? The intra-microaggregate material can be separated in micro aggregates and characterized by the range of techniques outlined in this chapter. Also, the light fraction can be determined, including roots and hyphae. The slow pool is a difficult problem and involves obtaining an estimate of OM in macroaggregates that is not within microaggregates. Perhaps this can be done by disruption of macroaggregates followed by removal of microaggregates. In such a scheme, organic materials associated within fine clay and on microaggregate surfaces remain a problem. However, there is currently evidence to suggest that such fractionation of OM may lead to quantitative assessment of pools with different turnover times, which can be used in simulation models. Alternately the information may lead to modification of current models.

Conclusions

1. It is clear that the turnover rate of SOM depends not only on the kinds of OM that are present but upon the position or location of OM in the soil.

2. There is a range of physical procedures that can be used to disaggregate soils from very gentle (e.g., controlled slow wetting and wet sieving) to very vigorous (e.g., use of ultrasonic energy).

Dynamics of SOM in Tropical Ecosystems

3. There is a range of particle separation methods available (e.g., sieving and sedimentation under a range of gravity forces and density methods.)
4. Chemical treatment of the soil before sonication or shaking can enhance disruption, and specific types of OM bonding agents can be targeted by using specific agents.
5. Application of physical methods to fractionate OM in tropical soils should be pursued to determine what pools can be measured and used in predictive models. It remains to be seen whether results obtained with tropical soils will be similar to temperate arable soils or if new conceptual models will need to be developed.

Methods of Characterization

A wide variety of chemical techniques have been used to characterize SOM. They include methods for carbohydrates (also polysaccharides), lipids, chemical and physicochemical properties of isolated humic substances (e.g., functional groups, molecular weights, and size and shapes of particles), and organic forms of N, P, and S. Details of these methods are available in the literature (e.g., Schnitzer, 1972; Schnitzer and Khan, 1978; Stevenson, 1982a; Aiken et al., 1985; Orlov, 1985; Kumada, 1987).

Spectroscopic techniques (e.g. visible and ultraviolet, infrared, nuclear magnetic resonance [NMR], and fluorescence) extend and supplement information available by chemical procedures (see above references). They serve as survey tools and provide a basis for selection of chemical techniques.

Emphasis will be given herein to two relatively new methods that show promise for characterization of OM in intact soils, namely, ^{13}C-NMR and pyrolysis/mass spectrometry (Py/Ms). Focus will be given to the types of information that can be obtained, the appropriateness of the methods to problems that can be encountered, and areas in which the methodology needs further development.

Whole Soil ^{13}C-NMR

Recent developments in solid-state NMR have greatly facilitated its application to OM in intact soils (as well as separated fractions). The technique of cross-polarization magic-angle spinning (CPMAS) was developed to overcome problems of broadening and low sensitivity in solids, and further improvements are possible using "dephasing," a variation of the basic CPMAS approach in which an extra delay period is used to enhance signal responses. Theory and methods of ^{13}C-NMR have been well-documented, and there are now several books and reviews on application of the method to SOM (Wershaw, 1985; Wilson 1981, 1987). Typical sample sizes are 100 to 300 mg.

The technique of CPMAS ^{13}C-NMR has been applied to mineral soil and its size and density fractions (Barron et al., 1980; Wilson et al., 1981a, 1981b,

1981c,; Skjemstad et. al., 1986, Oades et al., 1987), the humin fraction of SOM (Hatcher et al., 1980, 1985), Histosols (Preston and Ripmeester, 1982, 1983; Preston et al., 1984, 1987, 1989) and forest litter (Wilson et al., 1983; Ogner, 1985; Kogel et al., 1988). The technique has been used to follow changes in SOM due to cultivation (Skjemstad et al., 1986; Oades et al., 1987) and natural processes of decomposition (Kogel et al., 1988).

A CPMAS ^{13}C-NMR spectrum for a Mapourika soil from New Zealand is shown in Figure 4. Major ^{13}C resonances of humic substances and their assignments are as follows:

Alkyl (paraffinic) C	0-50 ppm
Aliphatic C-O, such as carbohydrates	
(~75 ppm) and methoxyl (~55 ppm)	50-100 ppm
Aromatic C (also alkenes)	100-160 ppm
Carboxyl C (includes COO-)	160-190 ppm
Carbonyl C = O (esters and amides)	190-200 ppm

The aromatic region consists of two groups of resonances, one near 128 ppm and the other at 150 ppm. The latter is characteristic of C next to O and has been assigned to phenolic C.

Results of ^{13}C-NMR studies provide approximate estimates for various types of organic C in SOM, namely, alkyl (paraffinic), aromatic, COOH,

Figure 4. ^{13}C-NMR spectrum of a Mapourika soil from New Zealand. See text for band assignments. (From Wilson et al., 1981c.)

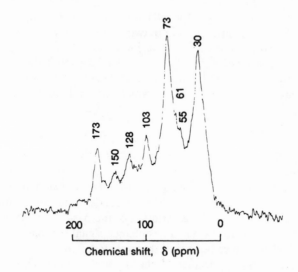

Chemical shift, δ (ppm)

Dynamics of SOM in Tropical Ecosystems

phenolic, aliphatic C-O or polysaccharide, and methoxyl. The fraction of the C accounted for as aromatic C has been referred to as the "aromaticity" of SOM, although in some work the C of COOH groups has been omitted when making the calculations (Hatcher et al., 1981; Schnitzer and Chan, 1986).

As with any other spectroscopic or analytical technique, results of CPMAS ^{13}C-NMR of organic materials must be interpreted with caution. While it is usual to analyze the spectra by dividing them into regions corresponding to chemical classes, these "chemical-shift" regions are not completely exclusive or specific. For SOM, processes of decomposition and humification may produce chemical structures that are no longer amenable to conventional interpretations. Many CPMAS ^{13}C-NMR spectra of SOM indicate a higher proportion of O-alkyl structures (carbohydrate C) than can be accounted for by conventional analyses for monosaccharides. Also, many spectra indicate a higher proportion of alkyl (or lipid-like) C that is not extractable by organic solvents. In general, NMR studies have indicated a lower phenolic, but higher COOH, content than chemical methods. Schnitzer and Chan (1986) suggested that, for humic substances, the signals for phenolic C are shifted downfield into the region of COOH groups and that resonances for the two overlap.

While the application of CPMAS ^{13}C-NMR to highly organic materials (i.e., Histosols, the litter layers of forest soils, and the light fraction of SOM) has been relatively easy and highly rewarding, there are serious problems in application of the technique to whole mineral soil or physical fractions. Due to a combination of relative low C and high Fe content, spectra of mineral soils are often broad and featureless. Recent work, however, has offered several approaches for obtaining more informative spectra of low-C samples. Thus, physical fractionations of soil can be used to concentrate C and chemical procedures can be used to dissolve inorganic components (Arshad et al., 1988; Preston et al., 1989).

Limitations and problems associated with CPMAS ^{13}C-NMR are summarized below.

1. Since many tropical soils are notoriously low in OM, special techniques will need to be developed in order for the method to be widely used.

2. A solid state ^{13}C-NMR spectrum represents the sum of all organic structures in the sample and does not distinguish between (a) humic substaces that vary in molecular weight and (b) structures that are associated with inorganic soil components (Skjemstad et al., 1986).

3. Costs for the purchase and maintenance of a modern ^{13}C-NMR spectrometer is beyond the means of most laboratories where research on SOM is underway. High priority must be given to establishment of effective working relationships between soil scientists involved in research on OM in tropical ecosystems and chemists with access to specialized equipment and instruments.

Pyrolysis/Mass Spectrometry (Py/Ms)

Analytical pyrolysis has had wide application in humus chemistry. As applied to intact soils, two approaches have been used, pyrolysis/gas chromatography (Py/GC) and Py/MS. A combination of the two has been applied to isolated components of SOM (Saiz-Jimenez and de Leeuw, 1986a). Advantages of Py/MS, in which the pyrolysis products are passed directly into a mass spectrometer, include better reproducibility, detection of highly polar species, fast analysis time, and amenability to multivariate or other chemometric methods of pattern analysis. Emphasis is given herein to Py/MS. Early work using Py/GC was reviewed by Bracewell and Robertson (1984).

In analytical pyrolysis, a pulse of thermal energy is applied to the sample, thereby causing fracture of weaker linkages of associated macromolecules, with release of products characteristic of their structures. Ratios obtained for products derived from polysaccharides, polypeptides, lignin, etc. can serve as a semiquantitative index of soil humus type and "degree of humification."

Two main pyrolysis techniques have been used: (1) quasi-instantaneous heating, or Curie-point pyrolysis, and (2) controlled temperature programming. A major criticism of the second technique is that secondary reactions during pyrolysis can lead to the formation of compounds unrelated to the material being pyrolyzed. Curie-point methods are nonselective and highly reproducible, and they can be applied to small sample sizes.

Curie-point pyrolysis methods have been used to distinguish between the OM of soils of different origins (Halma et al., 1978; Bracewell and Robertson, 1984, 1987b), to delineate the raw humus layers of forest soils (Bracewell and Robertson, 1987a), and to follow the humification process in peat (Bracewell et al., 1980b). Extensive use has also been made of the technique in structural studies of isolated fractions of SOM. The method has been proposed as a biomarker for terrestrial plant input into sediments (Saiz-Jimenez and de Leeuw, 1986b).

Some specific points relative to Curie-point pyrolysis are as follows:
1. Curie-point pyrolysis methods have the potential for differentiating between the different forms of SOM (i.e., lignin-like products, complex polysaccharides, proteinaceous constituents and lipids). In this respect, the approach has the potential for serving as a "fingerprint" technique for comparative purposes.

2. Due to the complexity of the mass spectra, a large number of variables (mass ions) must be reduced to a few principal component variables, which requires modern computer capability (Bracewell and Robertson, 1984). Selection of mass ions is somewhat arbitrary and has been based on fragmentation patterns of known biopolyers.

Pyrophosphate as an Extractant of Organically Bound Forms of Micronutrient Cations

The ability of complexing agents (such as EDTA or pyrophosphate) to dissolve SOM is well-known and these reagents have been used as selective extractants of organically bound forms of micronutrient cations. The most common reagent has been 0.1 M $K_4P_2O_7$.

Estimates for organically bound forms of micronutrient cations have generally been carried out in conjunction with a more extensive fractionation of the cations. In the scheme of McLaren and Crawford (1973), soluble-plus exchangeable-Cu was determined by neutral salt extraction ($CaCl_2$); specifically adsorbed Cu by extraction with dilute acetic acid; organically bound Cu by extraction with $K_4P_2O_7$; oxide occluded Cu by treatment with oxalate under ultraviolet light; and mineral lattice Cu by HF digestion of the final soil residue. For 24 contrasting soil types, from one-fifth to one-half of the Cu occurred in organically bound forms.

McLaren and Crawford (1973) concluded that the bulk of the "available" Cu in soils resided in organically bound forms and that the amount in forms available to plants (exchangeable- and soluble-Cu) was controlled by equilibria involving the specifically adsorbed and organically bound Cu, as follows:

$$\text{Exchangeable- and soluble-Cu} \longleftrightarrow \text{Specifically adsorbed Cu} \longleftrightarrow \text{Organically bound Cu}$$

Methods for determining organically bound forms of micronutrient cations in the soil solution were described by Stevenson and Fitch (1986).

The C/N/P/S Ratio

Some indication as to the quantity and quality of SOM can be obtained from analytical data for organic C, N, P, and S. Total OM can be estimated from the C value (OM = organic C x 1.727), although it should be noted that the conversion factor is not constant for all soils. Procedures for determining organic C, N, P, and S in soils are adequately described by Page et al. (1986).

Absolute amounts of organic N, P, and S vary greatly and are influenced by those factors that affect SOM levels, namely, the soil-forming factors of climate, topography, vegetation, parent material, and age. Organic matter content (and associated nutrients) usually declines when soils are first placed under cultivation and new equilibrium levels are attained characteristic of the cropping system employed. Factors affecting SOM levels have been covered in several reviews (Campbell, 1978; Stevenson, 1982a, 1986).

A definite relationship has been observed between organic C, total N, organic P, and total S in soils. Although variations exist in the C/N/P/S ratio for individual soils, the mean value for soils from different regions of the world is remarkably similar (Table 1): As an average, the proportion of C/N/P/S in soils is approximately 140:10:1.3:1.3.

Table 1. Organic C, total N, organic P, and total S relationship in soil.

Location*	Number of soils	C/N/P/S
Iowa, USA	6	110/10/1.4/1.3
Brazil	6	194/10/1.2/1.6
Scotland		
Calcareous	10	113/10/1.3/1.3
Noncalcareous	40	147/10/2.5/1.4
New Zealand	22	140/10/2.1/2.1
India	9	144/10/1.9/1.8

*For specific references see Stevenson (1986).

Native humus has a more consistent C/N/P/S ratio than crop residues, for the reason that decay leads to loss of a portion of the plant C as CO_2, with incorporation of nutrients into the microbial biomass (and hence into components of stable humus). The N, P, and S contents of microbial tissue vary somewhat but will be relatively constant for any given soil or for any group of related soils. Thus, the gradual transformation of plant raw material into stable OM (humus) leads to the establishment of a reasonably consistent relationship between C, N, P, and S. Changes in the C/N/P/S ratio of SOM, such as induced through cultivation, have been explained in terms of differences in the nature of organic N, P, and S (i.e., N and S occur as structural components of humic substances whereas P does not).

Traditionally, the C/N ratio has been assumed to be characteristic of SOM. This is not entirely true, for the reason that appreciable amounts of the N (usual range of 3 to 6% for surface soils) can occur as NH_4^+ held within the lattice structures of clay minerals.

Next to N, P is the most abundant nutrient contained in microbial tissue, making up as much as 2% of the dry weight. Largely for this reason, P is the second most abundant nutrient in SOM. The P content of SOM varies from as little as 1.0% to well over 3.0%, which is reflected by the variable C/organic P ratios that have been reported (range of 60 to 160 as deduced from the data in Table 1).

Ratios recorded for C/organic S are somewhat less variable than those for C/organic P, the usual range being 60 to 120. Difference in the C/S ratio between soil groups has been attributed to such factors as parent material and type of vegetative cover.

Dynamics of SOM in Tropical Ecosystems

Mineralizable N, P, and S

An important aspect of SOM dynamics in tropical ecosystems is the flow of N, P, and S through the various pools as influenced by soil management practices and changes in the soil environment. Such methods are discussed in Duxbury et al. (see Chapter 2). The possible use of C and N mineralization data, together with associated changes in the contents of C and N in the microbial biomass, to determine the amount of N contained in the labile pool of SOM has been investigated by K. Robertson et al. (1988).

A priori one might expect that N, P, and S would be released to plant-available forms in the same ratios in which they occur in OM. However, this has not always been the case. Differences in apparent mineralization rates for N, P, and S may be due to in part to their occurrence in different organic compounds and SOM fractions and to variations in the N, P, and S contents of applied plant residues (thereby leading to different mineralization-immobilization rates as noted in Duxbury et al. [see Chapter 2] and Sanchez et al. [see Chapter 5]).

Attempts have been made from time to time to relate availability of N, P, or S to a particular organic form of the nutrient. A disappointing aspect of these studies is that no given type (i.e., amino acids in the case of N) has been found to be superior to any other form as a source of N, P, or S for plants (Stevenson, 1986). Surprisingly, the distribution of the various organic N forms, as determined by fractionations based on acid hydrolysis, is somewhat similar for all soils and is relatively unaffected by crop management practices (Sowden et al., 1977).

Conclusions and Recommendations

Some general observations regarding humus as a reservoir and source of N, P, and S are as follows:

1. Because of the somewhat constant C/N/P/S ratio, the amounts of organic N, P, and S in soil can generally be estimated from OM content. However, unreliable values can be obtained for any given soil. Variations in C/P and C/S ratios are fully as great between a similar group of soils as for soils of contrasting soil types.

2. The assumption is often made that from 1 to 3% of the SOM is mineralized each year but the amounts liberated are compensated for by incorporation of similar amounts into newly formed humus. Thus, a net annual release of nutrients will only occur when SOM levels are declining.

3. When soils are cultivated regularly, availability of organic bound forms of the nutrients decrease, due in part to physical interactions with inorganic soil components and to changes in humus chemistry.

4. In many locations in the tropics, cropping systems are being used in which land is cleared, cropped for several years, and then returned to the native state for regeneration. Very little is known of the nutrient interchanges be-

tween the various OM pools in such ecosystems or of the conditions leading to optimum uptake of nutrients by plants.

5. When possible, agricultural practices should be adopted that will lead to maintenance of SOM at the highest level possible. The higher the return of crop or organic residues, the greater will be the magnitude of the cycling process and the greater will be the amounts of N, P, and S made available to the plant. Organic matter also has indirect effects on nutrient availability (including micronutrient cations), as discussed by Stevenson (1986).

6. Some reports indicate that components of the FA fraction of SOM constitute the biologically active fraction and transitory storehouse form of organically bound N, P, and S in soils, but this has yet to be confirmed. In a recent study with an ^{15}N-enriched Illinois Mollisol, He et al. (1988b) found that, of the mild extractants proposed as indexes of plant-available N, aqueous phenol showed the greatest promise as a selective extractant of potentially available (labile) organic N.

As far as research is concerned, the following recommendations can be made:

1. Expand research on methods for identifying those components of humus that are primarily involved in mineralization-immobilization turnover (so-called labile fractions). This information is required for the modeling of nutrient transformations and for predicting their release to available mineral forms during the course of the growing season.

2. Encourage use of isotopes to follow the interchange of nutrients between the various OM pools, including the so-called stable humus fraction. Double labeling should be used whenever possible (e.g., crop residues with ^{15}N and ^{13}C or ^{14}C). Results of ^{15}N studies with temperate-zone soils show that about one-third of the fertilizer N is retained in organic forms at the end of the first growing season. Only a small fraction (about 15%) of this residual N becomes available to plants during the second growing season and availability decreases even further in subsequent years. It is not known whether a similar sequence is followed in tropical soils.

Pools of Soil Organic Matter

The modern approach to complex natural systems is to develop a conceptual model in which the various "pools" are linked to each other in different ways. Ideally, the pools should have a real existence and be measurable so that fluxes (e.g. of C, N, P, and S) through the pools can be determined and thus the dynamics of the whole system monitored. It is clear that many of the chemical fractions outlined in this chapter are not conceptual pools but procedurally defined fractions and are of limited value in studies of the dynamics of SOM turnover. Conceptual pools currently used by modelers are described by Parton et al. (1987). They are based on turnover times as mentioned earlier, i.e. an active pool with a turnover time of 1.5 years, a slow pool with a turnover time of 25 years, and a passive pool with a turnover time of 1,000 years. The various pools of SOM that can currently be measured, albeit with difficulty, are considered below.

Litter

Litter is defined as the macroorganic matter that lies on the soil surface and is of considerable importance from the standpoint of nutrient cycling, particularly in forest soils and natural grasslands (see Sanchez et al., Chapter 5).

Herbaceous litter (including grasses and forest ground flora) is determined by harvesting known plot areas (usually 25 x 25 cm quadrants). Where possible, litter should be separated by plant species for the most frequent 80% of species and bulked for the remaining 20%, although in highly variable plant communities this may not be practical. Litter quantities should be expressed on an oven-dry basis in terms of $g\ m^{-2}yr^{-1}$ or $Mg\ ha^{-1}yr^{-1}$ with 95% confidence limits. Guidelines for quantification of litter inputs are given by J. M. Anderson and Ingram (1989).

Tree and shrub litter is usually collected using litter traps (usually 1 m in diameter) randomly located or in a stratified random pattern in sites where major variations exist in topography, soils, or vegetation type. Frequent collections are required for litters that decompose rapidly (such as some tree legumes); less frequent collections may be made for ecosystems under low rainfall (dry) conditions. It is often desirable to sort the material into *leaves* (including petioles and foliar rachises), *small woody litter* (twigs and bark), *reproductive structures* (flowers and fruit), and *trash* (sieve fraction less than 5 mm). Guidelines for litterfall measurements were given by Proctor (1983).

Light Fraction

The most common use of heavy liquids has been to separate organic materials not associated with mineral particles in what is often called the "light fraction." Densities used to separate light fractions have ranged from 1.6 to 2.0 $Mg\ m^{-3}$. Organic material separated at the lower density, consisting largely of plant debris and fauna, generally contains low amounts of inorganic materials, as determined as ash after ignition. There is no agreed division between obvious plant litter and the light fraction, although 2 mm has been suggested.

There is a need for standardization of the lower size limit for litter, which would then become the upper size limit for the light fraction. A lower size limit of 250 μm has been suggested for the light fraction but this has not always been adhered to. In some work, a light fraction has been separated by flotation in water and sieved out on a 250 μm sieve, in which case the term "macroorganic matter" has been used and is synonymous with the light fraction.

Separation of the light fraction can be achieved using a commercially available root-washing apparatus. Soil previously sieved through a 2-mm mesh screen is placed in a brass container fitted internally with baffles and with water jets set horizontally at an angle of 45° with respect to the container wall. When water pressure is applied, the soil is dispersed and the light fraction material, plus any suspended fine organic-mineral components, overflows through an opening at the apex of a cone centrally located within the apparatus. The light fraction is retained by a 0.25-mm sieve fitted to the bottom of the apparatus. For soils high in clay content, an overnight soaking in 10% sodium hexametaphosphate may be required to aid dispersion

The light fraction material collected on the 0.25-mm sieve is washed with water and filtered using a Büchner funnel to remove excess water. Roots are removed (often a difficult process) and the material is dried at 85°C. It is desirable to determine ash content (e.g., by combustion at 550°C in a muffle furnace) so that results can be normalized on an ash-free basis. Small amounts of mineral particles in the light fraction can strongly influene nutrient content. Contamination with mineral particles is a particularly serious problem with volcanic soils.

Input from roots can be appreciable. J. M. Anderson and Ingram (1989) describe a hierarchical approach to root studies, including quantification of the root biomass and guidelines for estimating total root production and total root C input into the soil (see Anderson et al., Chapter 4).

In the absence of a suitable apparatus, the light fraction can be separated by flotation. The density of the medium is critical and the method is difficult to standardize. Sollins et al. (1984) recovered debris from the surface layer of some forest soils by repeated flotation in an NaI soilution adjusted to sp. gr. 1.6g cm^{-1}. Variable proportions of the SOM are present in the light fraction (see Theng et al., Chapter1, and Duxbury et al., Chapter2).

Litter also falls in the category of macroorganic matter, but, as noted above, this material lies on the soil surface and is normally removed before soil sampling.

Soluble Organics

Recovery of the soil solution can be achieved by applying suction to a porous collector inserted into the soil. The porous sampler (i.e., ceramic, alumdum, fritted glass plate) is installed to the desired soil depth, a vacuum is developed within the sampler and pore water is drawn into the sample chamber through the porous section. The soil solution is recovered by disconnecting the sample chamber or by applying suction or pressure to the sampler.

Methods have also been described for recovery of the soil solution from field-moist soil under laboratory conditions. They include column miscible displacement, centrifucation with and without an immiscible liquid, and use of ceramic or plastic filters. The reader is referred to Elkhatib et al. (1986) and Wolt and Graveel (1986) for reviews and evaluation of methods.

Microbial Biomass

The importance of the soil microbial biomass in the cycling of soil nutrients has been well-established and many methods exist for the determination of biomass C (as well as biomass N, P, and S). Most current approaches are based on the chloroform ($CHCl_3$) fumigation technique initially developed by Jenkinson and Powlson (1976).

Biomass C and N by Fumigation-Incubation

Fumigation of a soil leads to an increase in respiration rate with release of CO_2 and nutrients. Fumigation-incubation methods for biomass C and N are based on the assumption that the extra CO_2 and N released when a soil is fumigated and subsequently incubated comes from the cells of microorganisms killed by fumigation and decomposed by recolonizing microorganisms.

Biomass C is calculated from the relationship: Biomass C $= F_c/k_c$, where F_c is the difference between the amount of CO_2-C released by incubation of the fumigated and unfumigated soil and k_c is the fraction of the biomass C mineralized to CO_2 over the incubated period (usually 10 to 14 days).

The value chosen for k_c is of some importance and will not be precisely the same for all soils. However, k_c values obtained for pure cultures of bacteria and fungi span a rather narrow range, with most values falling within the range of 0.43 to 0.50 (J. P. E. Anderson and Domsch, 1980). For biomass C calculations, a k_c value of 0.45 is commonly used (Jenkinson and Ladd, 1981).

Assumptions underlying the $CHCl_3$ fumigation-incubation method for biomass C are: (1) the C in dead organisms is more rapidly mineralized than that in living organisms; (2) all microorganisms are killed during fumigation; (3) the death of organisms in the control soil is neglibible as compared to that in the fumigated soil; and (4) microbial decomposition of native SOM occurs at the same rate in fumigated and unfumigated soil (Jenkinson and Ladd, 1981). The last assumption is not valid for all soils, such as those receiving large inputs of fresh substrates (Jenkinson and Powlson, 1976; Martens, 1985).

The fumigation-incubation method is unsuitable for analysis of air-dried soils. Alternate methods for the determination of biomass C are discussed by Jenkinson and Ladd (1981) and Stevenson (1986, Chapter 1).

Results for biomass C have been used for estimating biomass N and P. In the approach of J. P. E. Anderson and Domsch (1980), data for biomass C were converted to nutrient values using measured C/N and C/P ratios for 24 pure cutlures of soil microorganisms and the relative contribution of bacterial and fungal cells to the microbial biomass.

For biomass N, measurements are made for the amount of N (mineral + organic) released after incubation of $CHCl_3$-fumigated soil and recovered by extraction with $0.5\,M$ K_2SO_4 (Jenkinson and Powlson, 1976; Shen et al., 1984). Biomass N is obtained from the relationship: Biomass $N = F_n/k_n$, where F_n is the difference between the amount of N released by incubation of fumigated and unfumigated soil and k_n is the fraction of the biomass N that is released to soluble forms during incubation. In contrast to k_c, estimates for k_n have been highly variable (Brookes et al., 1985). Based on the measured relationship between F_c and F_n, and the likely C to N ratios of the microbial biomass, a value for k_n of 0.68 has been recommended (Shen et al., 1984). However, this value must be used with caution on soils amended with large quantities of OM and not at all on strongly acid soils ($<$ pH 4.5).

A major problem with the fumigation-incubation method for estimating biomass N is that denitrification or immobilization may alter the amount of N accounted for in soluble forms after incubation. As a general rule, from 1 to 6% of the soil organic N resides in the microbial biomass at any one time.

Extraction Methods for Biomass C and N

In addition to the above, extraction methods have been described for the direct determination of biomass C (Sparling and West, 1988a; Tate et al., 1988), and N (Brookes et al., 1985). The basis for these methods is that the amounts of organic C and N in extracts of $CHCl_3$-fumigated soils are reasonably well-correlated with the corresponding amounts released to CO_2 or soluble N forms by incubation. In each case, the reagent used for extraction is $0.5\,MK_2SO_4$.

For reasons discussed by Sparling and West (1988a) and Tate et. al. (1988), conversion factors for estimating biomass C from extracted C values are highly variable; for mineral soils, a provisional value of 0.33 has been recommended. In the case of biomass N, a 1:1 relationship was observed by Brookes et al. (1985) for total N released by $CHCl_3$ after a 5-day fumigation period and the flush of mineral N as obtained by the fumigation-incubation method. Accordingly, the same conversion factors applies (i.e., $k_n = 0.68$).

Advantages of the modified approaches are that there is no need for complete removal of fumigant or for prolonged incubation of the soil under carefully controlled conditions. For biomass N, an additional advantage is that errors do not arise due to denitrification and immobilization of inorganic N during incubation.

Biomass P and S

Fumigation-extraction procedures similar to those noted above have been used for the determination of biomass P and biomass S. For biomass P, the extractant is $0.5\,M$ $NaHCO_3$ at pH 8.5 ($k_p = 0.4$) and requires a correction for phosphate fixation during extraction (Brookes et al., 1982; Hedley and Stewart, 1982). Two extractants have been used for biomass S, $0.1\,M$ $NaHCO_3$ and $0.01\,M$ $CaCl_2$, with k_s values of 0.41 and 0.35, respectively (Saggar et al., 1981).

Stable Humus

Physical and chemical procedures have been used in attempts to fractionate SOM and to determine the location of individual components within the soil matrix. These methods were described earlier.

Conclusions and Recommendations

Theme Imperative: Expand research on the chemistry and reactions of humus in tropical soils, including location of components within the soil matrix as influenced by soil management practices, quantification of labile fractions by physical/chemical methods and interchange of nutrients (N, P, and S) between stable humus and the other OM pools (i.e., litter, light fraction, and biomass), for which isotopes will need to be used.

Research Imperatives:

1. Develop reliable procedures for measurement of SOM pools, including litter, light fraction, microbial biomass (C, N, P, and S), and the labile and passive components of soil humus.

2. Reconcile differences among conceptual, mathematical, and operational definitions of SOM pools.

3. Improve understanding as to how spatial arrangement of organic substrates affects SOM dynamics (i.e., is physical protection more important than chemical protection in the slow turnover of stabilized OM).

4. Improve chemical procedures for extractions and fractionation of SOM (emphasis on biologically meaningful fractions).

5. Determine dynamics of SOM transformations in terms of nutrient fluxes (rather than static pool sizes). Promote more extensive use of stable and radioactive isotopes.

6. Promote increased research designed to clarify the role of OM in the cycling of micronutrient cations in tropical soils.

7. Encourage and promote cooperative research between soil scientists of tropical regions and their counterparts at research institution where specialized equipment is available, notably isotope and CPMAS ^{13}C-NMR capability.

Literature Cited

Abboud, S. A. and Turchenek, L. W. 1986. Solid state ^{13}C NMR spectroscopy of size and density fractions of a chernozemic soil. (Unpublished report)

Adams, W. A. and Kassim, J. K. 1984. Iron oxyhydroxides in soils developed from Lower Paleozoic sedimentary rocks in mid-Wales and implications for some pedogenetic process. Journal of Soil Science 35:117-126.

Ahmed, B. and Islam, A. 1978. Occurrence of inositol hexaphosphate in some humid tropical soils. Tropical Agriculture (Trinidad) 55:149-152.

Ahmed, M. and Oades, J. M. 1984. Distribution of organic matter and adenosine-triphosphate after fractionation of soils by physical procedures. Soil Biology and Biochemistry 16:465-470.

Aiken, G. R. 1988. A critical evaluation of the use of macroporous resins for the isolation of aquatic humic substances. In: F. H. Frimmel and R. F. Christman (eds.), Humic substances and their role in the environment. Wiley, New York. pp. 15-28.

Aiken, G. R., McKnight, D. M., Wershaw, R. L., and MacCarthy, P. (eds.) 1985. Humic substances in soil, sediment, and water. Wiley, New York.

Alegre, J. C. and Cassel, D. K. 1986. Effect of land-clearing methods and post-clearing management on aggregate stability and organic carbon content of a soil in the humid tropics. Soil Science 142:289-295.

Allen, J. C. 1985. Soil responses to forest clearing in the United States and the tropics: geological and biological factors. Biotropica 17:15-27.

Allison, F. E. 1973. Organic matter and its role in crop production. Elsevier, Amsterdam. 637 pp.

Almendros, G. and Gonzalez-Villa, F. J. 1987. Degradative studies on a soil humin fraction — sequential degradation of inherited humin. Soil Biology and Biochemistry 19:513-520.

Almendros, G., Martin, F., and Gonzalez-Villa, F. J. 1988. Effects of fire on humic and lipid fractions in a Dystric Xerochrept in Spain. Geoderma 42:115-127.

Amato, M., Ladd, J. N., Ellington, A., Ford, G., Mahoney, J. E., Taylor, A. C., and Walsgott, D. 1987. Decomposition of plant material in Australian soils. IV. Decomposition in situ of ^{14}C and ^{15}N-labelled legume and wheat materials in a range of southern Australian Soils. Australian Journal of Soil Research 25:95-105.

Anderson, D. W. 1979. Processes of humus formation and transformation in soils of the Canadian Great Plains. Journal of Soil Science 30:77-84.

Anderson, D. W. and Paul, E. A. 1984. Organo-mineral complexes and their study by radiocarbon dating. Soil Science Society of America Journal 48:298-301.

Anderson, D. W., Saggar, S., Bettany, J. R., and Stewart, J. W. B. 1981. Particle size fractions and their use in studies of soil organic matter. I. The nature and distribution of forms of carbon, nitrogen, and sulfur. Soil Science Society of America Journal 45:767-772.

Anderson, G. 1980. Assessing organic phosphorus in soils. In: F. E. Khasawneh, E. C. Sample, and E. J. Kampreth (eds.), The role of phosphorus in agriculture. American Society of Agronomy, Madison, Wisconsin.

Anderson, G. and Arlidge, E. Z. 1962. The adsorption of inositol phosphates and glycerophosphate by soil clays, clay minerals, and hydrated sesquioxides in acid media. Journal of Soil Science 13:216-224.

Anderson, G., Williams, E. G., and Moir, J. O. 1974. A comparison of the sorption of inorganic orthophosphate and inositol hexaphosphate by six acid soils. Journal of Soil Science 25:51-62.

Anderson, J. M. 1988a. Fauna mediated transport processes in soils. Agriculture, ecosystems and environment 24: 5-19.

————. 1988b. Spatiotemporal effects of invertebrates on soil processes. Biology and Fertility of Soils 6:216-227.

Anderson, J. M. and Ingram, J. S. I. (eds.) 1989. Tropical soil biology and fertility handbook of methods. Commonwealth Agricultural Bureau International, Wallingford, England.

Anderson, J. M. and Swift, M. J. 1983. Decomposition in tropical forests. In: S. L. Sutton, T. C. Whitmore, and A. C. Chadwick (eds.), Tropical rainforests: ecology and management. Special Publication 2, British Ecological Society. Blackwell Scientific, Oxford, England. pp. 287-309.

Anderson, J. M. and Wood, T. G. 1984. Mound composition and soil modification by two soil-feeding termites (Termitinae, Termitidae). Pedobiologia 26:77-82.

Anderson, J. M., Proctor, J., and Vallack, H. W. 1983. Ecological studies in four contrasting lowland rainforests in Gunung Mulu National Park, Sarawak. III. Decomposition processes and nutrient losses from leaf litter. Journal of Ecology 71:503-527.

Anderson, J. P. E. and Domsch, K. H. 1980. Quantities of plant nutrients in the microbial biomass of selected soils. Soil Science 130:211-216.

Andren, O. and Lagerlöf, J. 1983. Soil fauna (microarthropods, enchytraeids, nematodes) in Swedish agricultural cropping systems. Acta Agric. Scand. 33:33-52.

Arshad, M. A., Ripmeester, J. A., and Schnitzer, M. 1988. Attempts to improve solid state ^{13}C NMR spectra of whole soils. Canadian Journal of Soil Science 68:593-602.

Atlas of Hawaii, 2nd ed. 1983. Department of Geography, University of Hawaii. University of Hawaii Press, Honolulu.

Audreux, F., Faivre, P., and Bonne, M. 1987. Nature et dynamique des matières organiques dans les processes de podzolisation. In: D. Righi and A. Chauvel (eds.), Podzols et podzolisation. Institut National de la Recherche Agronomique, Paris. pp. 119-130.

Audreux, F., Faivre, P., and Ruiz, E. 1985. Influence du gradient climatique sur la differenciation des sols d'une séquence altitudinale dans les Andes de Colombie. Distribution des matières humiques et rôle du fer dans l'humification. Comptes Rendus de l'Académie des Sciences, Paris, Série II. 300:223-226.

Avnimelech, Y. 1986. Organic residues in agriculture. In: Y. Chen and Y. Avnimelech (eds.), The role of organic matter in modern agriculture. Martinus Nijhoff, Dordrecht, The Netherlands. pp. 1-10.

Ayanaba, A., Tuckwell, S. B., and Jenkinson, D. S. 1976. The effects of clearing and cropping on the organic reserves and biomass of tropical forest soils. Soil Biology and Biochemistry 8:519-525.

Azhar, El Sayed, Verhe, R., Proot, M., Sandra, P., and Verstraete, W. 1986. Binding of nitrite-N on polyphenols during nitrification. Plant and Soil 94:369-382.

Balesdent, J., Mariotti, A., and Guillet, B. 1987. Natural carbon-13 abundance as a tracer for studies of soil organic matter dynamics. Soil Biology and Biochemistry 19:25-30.

Balesdent, J., Wagner, G. H., and Mariotti, A. 1988. Soil organic matter turnover in long-term field experiments as revealed by carbon-13 natural abundance. Soil Science Society of America Journal 52:118-124.

Barber, D. A. and Martin, J. K. 1976. The release of organic substances by cereal roots into soil. New Phytologist 76:69-80.

Barber, S. A. 1984. Soil nutrient bioavailability. Wiley, New York.

———. 1979. Corn residue management and soil organic matter. Agronomy Journal 71:625-627.

Barois, I. and Lavelle, P. 1986. Changes in respiration rate and some physicochemical properties of a tropical soil during transit through *Pontoscolex corethrurus* (Glossoscolecidae, Oligochaeta). Soil Biology and Biochemistry 18:539-541.

Barois, I., Cadet, P., Albrecht, A., and Lavelle, P. 1988. Systemes de culture et faune des sols. In: C. Feller (ed.), Fertilite des sols dans les agricultures paysannes caribeennes. Effet des restitutions organiques. Projet CEE/ORSTOM no. TSDA-0l78F. pp. 85-94.

Barron, P. F., Wilson, M.A., Stephens, J. F., Cornell, B. A., and Tate, K. R. 1980. Cross polarisation ^{13}C n.m.r. spectroscopy of whole soil. Nature. (London) 286:585.

Barrow, N. J. 1961. Phosphorus in soil organic matter. Soils and Fertilizer 24:169-173.

Beauchamp, E. G., Reynolds, W. D., Brasche-Villeneuve, D., and Kirby, K. 1986. Nitrogen mineralization kinetics with different soil pretreatments and cropping histories. Soil Science Society of America Journal 50:1478-1483.

Benner, R., Fogel, M. L., Sprague, E. K., and Hodson, R. E. 1987. Depletion of carbon-13 in lignin and its implications for stable carbon isotope studies. Nature 329:708-710.

Benoit, R. E. and Starkey, R. L. 1968. Inhibition of decomposition of cellulose and some other carbohydrates by tannin. Soil Science 105:291-296.

Berendse, F., Berg, B., and Bossata, E. 1987. The effect of lignin and nitrogen on the decomposition of litter in nutrient-poor ecosystems: a theoretical approach. Canadian Journal of Botany 65:1116-1120.

Berg, B., Muller, M. and Wessen, B. 1987. Decomposition of red clover (*Trifolium pratense*) roots. Soil Biology and Biochemistry 19:589-593.

Berish, C. W. 1982. Root biomass and surface area in three successional tropical forests. Canadian Journal of Forestry Research 12:699-704.

Berish, C. W. and Ewel, J. J. 1988. Root development in simple and complex tropical successional ecosystems. Plant and Soil 106:73-84.

Bertrand, A. R. 1983. The key to the agricultural ecosystem. In: R. R. Lowrance, R. L. Todd, L. E. Asmussen, and R. A. Leonard (eds.), Nutrient cycling in agricultural ecosystems. Special Publication 23. University of Georgia Agricultural Experiment Station, Athens. pp. 3-12.

Bhatti, H. M., Yasin, M., and Rashad, M. 1985. Evaluation of Sesbania green manuring in rice-wheat rotation. In: Proceedings of the International Symposium on Nitrogen and the Environment. NIAB, Faisalabad, Pakistan. pp. 275-284.

Biederbeck, V. O. 1978. Soil organic matter and soil fertility. In: M. Schnitzer and S. U. Khan (eds.), Soil organic matter. Elsevier, New York. pp. 273-310.

Bignell, D. E., Oskarsson, H., Anderson, J. M., Ineson, P., and Wood, T. G. 1983. Structure, microbial associations and function of the so-called "mixed segment" of the gut in two soil-feeding termites, *Procubitermes aburiensis* and *Cubitermes*

Dynamics of SOM in Tropical Ecosystems

severus (Termitinae, termitidae). Journal of the Zoological Society of London 201:445-480.

Billes, G. and Bottner, P. 1981. Effet des racines vivantes sur la décomposition d'une litiere racinaire marquée au ^{14}C. Plant and Soil 62:193-208.

Birch, H. F. 1958. The effect of soil drying on humus decomposition and nitrogen availability. Plant and Soil 10:9-31.

Bonde, T. A. and Lindberg, T. 1988. Nitrogen mineralization kinetics in soil during long-term aerobic laboratory incubations: a case study. Journal of Environmental Quality 17:414-417.

Boone, R. D., Sollins, P., and Cromack, K., Jr. 1988. Stand and soil changes along a mountain hemlock death and regrowth sequence. Ecology 69:714-722.

Bornemisza, E. and Igue, K. 1967. Comparisons of three methods for determining organic phosphorus in Costa Rican soils. Soil Science 103:347-353.

Bouché, M. B. 1977. Strategies lombriciennes. In: U. Lohm and T. Persson (eds.), Soil organisms as components of ecosystems. Ecological Bulletins (Stockholm) 33:122-132.

Bouldin, D. R. 1988. The effect of green manure on soil organic matter content and N availability to crops. IRRI, Los Banos, The Philippines. In Press.

Bowen, G. D. 1984. Tree roots and the use of soil nutrients. In: G. D. Bowen and E. K. S. Nambiar (eds.), Nutrition of plantation forests. Academic Press, London. pp. 147-479.

Bowen, G. D. and Nambiar, E. K. S. (eds.) 1984. Nutrition of plantation forests. Academic Press, New York. 516 pp.

Bracewell, J. M. and Robertson, G. W. 1984. Characteristics of soil organic matter in temperate soils by Curie-point pyrolysis-mass spectrometry. I. Organic matter variations with drainage and mull humification in A horizons. Journal of Soil Science 35:549-558.

Bracewell, J. M. and Robertson, G. W. 1987a. Characteristics of soil organic matter in temperate soils by Curie-point pyrolysis-mass spectrometry. II. The effect of drainage and illuviation in B horizons. Journal of Soil Science 38:181-198.

Bracewell, J. M. and Robertson, G. W. 1987b. Characteristics of soil organic matter in temperate soils by Curie-point pyrolysis-mass spectrometry. III. Transformations occurring in surface organic horizons. Geoderma 40:333-344.

Bracewell, J. M. and Robertson, G. W. 1987c. Indications from analytical pyrolysis on the evolution of organic materials in the temperate environment. Journal of Analytical and Applied Pyrolysis 11:355-366.

Bracewell, J. M., Robertson, G. W., and Tate, K. R. 1976. Pyrolysis-gas chromatography studies on a climosequence of soils in tussock grasslands, New Zealand. Geoderma 15:209-215.

Bracewell, J. M., Robertson, G. W., and Welch, D. I. 1980a. Polycarboxylic acids as the origin of some pyrolysis products characteristic of soil organic matter. Journal of Analytical and Applied Pyrolysis 2:239-248.

Bracewell, J. M., Robertson, G. W., and Williams, B. L. 1980b. Pyrolysis-mass spectrometry studies of humification in a peat and a peaty podzol. Journal of Analytical and Applied Pyrolysis 2:53-62.

Brookes, P. C., Powlson, D. S., and Jenkinson, D. S. 1982. Measurement of microbial biomass phosphorus in soil. Soil Biology and Biochemistry 14:319-329.

Brookes, P. C., Landman, A., Pruden, G., and Jenkinson, D. S. 1985. Chloroform fumigation and the release of soil nitrogen: a rapid direct extraction method to measure microbial biomass nitrogen in soil. Soil Biology and Biochemistry 17:835-842.

Brown, S. and Lugo, A. E. 1982. The storage and production of organic matter in tropical forests and their role in the global carbon cycle. Biotropica 14:161-187.

Buschbacher, Uhl, C. and Searrao, E. A. S. 1988. Abandoned pastures in eastern Amazonia. II. Nutrient stocks in the soil and vegetation. Journal of Ecology 76:682-699.

Buxton, R. D. 1981. Changes in the composition and activities of termite communities in relation to changing rainfall. Oecologia (Berlin) 51:371-378.

Cabala-Rosand, P. (ed.) 1985. Reciclagen de nutrientes e agricultura de baixos insumos nos trópicos. Centro de Pesquisas do Cacau, Ilheus, Bahía, Brazil. 341 pp.

Cabrera, M. L. and Kissel, D. E. 1988a. Potentially mineralizable nitrogen in disturbed and undisturbed soil samples. Soil Science Society of America Journal 52:1010-1015.

Cabrera, M. L. and Kissel, D. E. 1988b. Evaluation of a method to predict nitrogen mineralized from soil organic matter under field conditions. Soil Science Society of America Journal 52:1027-1031.

Caldwell, M. M. 1979. Root structure: the considerable cost of below-ground function. In: O. T. Solbrig, S. Jain, G. B. Johnson, and P. H. Raven (eds.), Topics in plant population biology. Columbia University Press, New York. pp. 408-430.

Cameron, R. S. and Posner, A. M. 1979. Mineralizable organic nitrogen in soil fractionated according to particle size. Journal of Soil Science 30:565-577.

Campbell, C. A. 1978. Soil organic carbon, nitrogen and fertility. In: M. Schnitzer and S. U. Khan (eds.), Soil organic matter. Elsevier, New York. pp. 173-271.

Campbell, C. A., Paul, E. A., Rennie, D. A., and McCallum, K. J. 1967. Applicability of the carbon-dating method of analysis to soil humus studies. Soil Science 104:217-224.

La Caro, F. and Rudd, R. L. 1985. Leaf litter disappearence rates in Puerto Rican montane rain forest. Biotropica 17:269-276.

Carter, M. R. and Rennie, D. A. 1987. Effects of tillage on deposition and utilization of ^{15}N residual fertilizer. Soil Tillage Research 9:33-43.

Catroux, G. and Schnitzer, M. 1987. Chemical, spectroscopic, and biological characteristics of the organic matter in particle size fractions separated from an Aquoll. Soil Science Society of America Journal 15:1200-1207.

Cavallaro, N. and McBride, M. B. 1984. Effect of selective dissolution on charge and surface properties of an acid soil clay. Clays and Clay Minerals 32:283-290.

CEPLAC. 1985. Informe de pesquisas de 1983. Comissáo Executiva do Plano de Laxoura Cacaueria, Ilheus, Bahía, Brazil.

Cerri C. C., Feller, C., Balesdent, J., Victoria, R., and Plenecassagne, A. 1985. Application du traçage isotopique naturel en ^{13}C, a l'étude de la dynamique de la matière organique dan les sols. Comptes Rendus de l'Académie des Sciences, Paris, Série II. 300:423-428.

Chapin, F. S. III. 1980. The mineral nutrition of wild plants. Annual Review of Ecology and Systematics 11:233-260.

Chapin, F. S. and Kedrowski, R. A. 1983. Seasonal changes in nitrogen and phosphorus fractions and autumn retranslocation in evergreen and deciduous taiga trees. Ecology 64:376-391.

Charley, J. L. and Richards, B. N. 1983. Nutrient allocation in plant communities: mineral cycling in terrestrial ecosystems. In: O. L. Lange, P. S. Nobel, C. B. Osmond, and H. Ziegler (eds.), Physiological plant ecology. IV. Ecosystem processes. Springer Verlag, Berlin. pp. 5-45.

Chatterjee, B. N., Singh, K. I., Pal, A., and Maiti, S. 1979. Organic manures as substitutes for chemical fertilizers for high yielding rice varieties. Indian Journal of Agricultural Sciences 49:188-192.

Cheshire, M. V., Mundie, C. M., and Shepherd, H. 1974. Transformation of sugars when rye hemicellulose labelled with ^{14}C decomposes in soil. Journal of Soil Science 25:90-95.

Cheshire, M. V., Sparling, G. P., and Mundie, C. M. 1985. The effect of oxidation by periodate on soil carbohydrate derived from plants and microorganisms. Journal of Soil Science 36:351-356.

Cheshire, M. V., Sparling, G. P., and Mundie, C. M. 1984. Influence of soil type, crop and air drying on residual carbohydrate content and aggregate stability after treatment with periodate and tetraborate. Plant and Soil 76:339-347.

Cheshire, M. V., Sparling, G. P. and Mundie, C. M. 1983. Effect of periodate treatment of soil and carbohydrate constituents and soil aggregation. Journal of Soil Science 34:105-112.

Chichester, F. W. 1969. Nitrogen in soil organic-mineral sedimentation fractions. Soil Science 107:356-363.

Christensen, B. T. 1987. Decomposability of organic matter in particle size fractions from field soils with straw incorporation. Soil Biology and Biochemistry 19:429-435.

Christensen, B. T.. 1986. Straw incorporation and soil organic matter in macro-aggregates and particle size separates. Journal of Soil Science 37:125-135.

Christensen, B. T. and Sørensen, L. H. 1986. Nitrogen in particle size fractions of soils incubated for five years with ^{15}N-ammonium and ^{14}C-hemicellulose. Journal of Soil Science 37:241-247.

Christensen, B. T. 1985. Carbon and nitrogen in particle size fraction isolated from Danish arable soils by ultrasonic dispersion and gravity-sedimentation. Acta Agric. Scand. 35:175-187.

Christensen, B. T. and Sørensen, L. H. 1985. The distribution of native and labelled carbon between soil particle size fractions isolated from long-term incubation experiments. Journal of Soil Science 36:219-229.

Churchman, G. J. and Tate, K. R. 1986. Aggregation of clay in six New Zealand soil types as measured by disaggregation procedures. Geoderma 37:207-220.

CIAT. 1985. Tropical pastures program, annual report for 1984. Centro Internacional de Agricultura Tropical, Cali, Colombia. 137 pp.

Cline, G. R., Powell, P. E., Szaniszlo, P. J., and Reid, C. P. P. 1983. Comparison of the abilities of hydroxamic and other natural organic acids to chelate iron and other ions in soil. Soil Science 136:145-157.

Cole, C. V., Stewart, J. W. B., Ojima, D., Parton, W. J., and Schimel, D. 1989. Modeling land use effects on soil organic matter dynamics in the North American Great Plains. Plant and Soil. In press.

Coley, P. D. 1983. Herbivory and defensive characteristics of tree species in a lowland tropical forest. Ecological Monographs 53:209-233.

Collins, N. M. 1983. The utilization of nitrogen resources by termites (Isoptera). In: J. A. Lee, S. McNeill, and I. H. Rorison (eds.), Nitrogen as an ecological factor. Blackwell Scientific, Oxford, England. pp. 381-412.

Collins, N. M., Anderson, J. M., and Vallack, H. W. 1984. Studies on the soil invertebrates of lowland and montane rainforests in the Gunung Mulu National Park. Sarawak Museum Journal 30: 19-33.

Cooke, G. W. 1977. The roles of organic manures and organic matter in managing soils for higher crop yields – a review of the experimental evidence. Proceedings of the International Seminar on Soil Environment and Fertility Management in Intensive Agriculture, Tokyo, Japan.

Cotching, W. E., Allbrook, R. F., and Gibbs, H. S. 1979. Influence of maize cropping on the soil structure of two soils in the Waikato district, New Zealand. New Zealand Journal of Agricultural Research 22:431-438.

Couto, W., Lathwell, D. J., and Bouldin, D. R. 1979. Sulfate sorption by two Oxisols and an Alfisol of the tropics. Soil Science 127:108-116.

Crampton, C. B. 1982. Podsolization of soils under individual tree canopies in southwestern British Columbia. Geoderma 28:57-61.

Crossley, D. A. Jr., House, G. J., Snider, R. M., Snider, R. J., and Stinner, B. R. 1984. The positive interactions in agroecosystems. In: R. R. Lowrance, B. R. Stinner, and G. J. House (eds.), Agricultural ecosystems, unifying concepts. Wiley, New York. 233 pp.

Cuevas, E. 1983. Crecimiento de raices finas y su relación con los procesos de descomposición de materia orgánica y liberación de nutrientes en bosques del Alto Rio Negro en el Territorio Federal Amazonas. Ph.D. dissertation. Instituto Venezolano de Investigaciones Científicas, Caracas, Venezuela.

Cuevas, E. and Medina, E. 1988. Nutrient dynamics within Amazonian forests. II. Fine root growth, nutrient availability and leaf litter decomposition. Oecologia (Berlin) 76:222-235.

Cuevas, E. and Medina, E. 1986. Nutrient dynamics within Amazonian forest ecosystems. I. Nutrient flux in fine litter fall and efficiency of nutrient utilization. Oecologia (Berlin) 68:466-472.

Cuevas, E. and Medina, E. 1983. Root production and organic matter decomposition in a Tierra Firme forest of the Upper Rio Negro basin. In: L. Kutschere (ed.), Wurzelökologie und ihre Nutzanwendung (Root ecology and its practical applications). International Symposium on Root Ecology and Its Applications, Gumpenstein, Austria, 1982. pp. 653-666.

Dalal, R. C. and Mayer, R. J. 1986a. Long-term trends in fertility of soils under continuous cultivation and cereal cropping in southern Queensland. III. Distribution and kinetics of soil organic carbon in particle-size fractions. Australian Journal of Soil Research 24: 293-300.

Dalal, R. C. and Mayer, R. J. 1986b. Long-term trend in fertility of soils under continuous cultivaton and cereal cropping in southern Queensland. IV. Loss of organic carbon from different density fractions. Australian Journal of Soil Research 24: 301-309.

Darici, C., Herrera, V., and Schaefer, R. 1988. L'effet de divers tanins sur l'activite microbienne des sols tropicaux a allophane (Andisols) et a montmorillonite (Vertisols). Proceedings 113e Congrès Nationale de Sociétiés Savantes. In Press.

Dash, M. C., Behera, N., Satpathy, B., and Pati, D. P. 1985. Comparison of two methods for microbial biomass estimation in some tropical soils. Revue d'-Ecologie et de Biologie du Sol 22:13-20.

Davidson, E. A., Galloway, L. F., and Strand, M. K. 1987. Assessing available carbon: comparison of techniques across selected forest soils. Communications in Soil Science and Plant Analysis 18:45-64.

Deans, J. R., Molina, J. A. E., and Clapp, C. E. 1986. Models for predicting potentially mineralizable N and decomposition rate constants. Soil Science Society of America Journal 50:323-327.

DeDatta, S. K. 1981. Principles and practices of rice production. Wiley, New York. 618 pp.

DeHaan, S. 1976. Humus, its formation, its relation with the mineral part of the soil and its significance for soil productivity. In: Soil organic matter studies, vol. I. International Atomic Energy Agency, Vienna. pp. 21-30.

Dietz, S. and Bottner, P. 1981. Etude par autoradiographie de l'enfouissement d'une litière marquée au ^{14}C en milieu herbacé. In: Migrations organo-minérales dans les sols tempérés, Nancy, 1979. Centre National de la Recherche Scientifique, Paris. pp. 125-132.

Doran, J. W. 1987a. Microbial biomass and mineralizable nitrogen distributions in no-tillage and plowed soils. Biology and Fertility of Soils 5:68-75.

———. 1987b. Tillage effects on microbiological release of soil organic nitrogen. In: T. J. Gerik and B. L. Harris (eds.), Conservation tillage. Southern Region No-till Conference, July 1987. Texas A & M University, College Station. pp. 63-66.

Doran, J. W. and Smith, M. S. 1987. Organic matter management and utilization of soil and fertilizer nutrients. Chapter 4, in: R. F. Follett, J. W. B. Stewart, and C. V. Cole (eds.), Soil fertility and organic matter as critical components of production systems. Special Publication 19. Soil Science Society of America, Madison, Wisconsin. pp. 53-72.

Dormaar, J. F. 1984. Monosaccharides in hydrolysates of water-stable aggregates after 67 years of cropping to spring wheat as determined by capillary gas chromatography. Canadian Journal of Soil Science 64:647-656.

Dormaar, J. F. 1983. Chemical properties of soil and water-stable aggregates after sixty-seven years of cropping to spring wheat. Plant and Soil 75:51-61.

Douglas, J. T. and Goss, M. J. 1982. Stability and organic matter content of surface soil aggregates under different methods of cultivation and in grassland. Soil Tillage Research 2:155-175.

Duchaufour, P. 1977. Pedology: pedogenesis and classification. Allen and Unwin, London.

Dudal, R. 1976. Inventory of the major soils of the world with special reference to mineral stress hazards. In: M. J. Wright (ed.), Plant adaptation to mineral stress in problem soils. Cornell University Agricultural Experiment Station, Ithaca, New York.

Edwards, A. P. and Bremner, J. M. 1967. Microaggregates in soils. Journal of Soil Science 18:64-73.

Edwards and Grubb. 1982. Studies of mineral cycling in a montane rain forest in New Guinea. IV. Soil characteristics and the division of mineral elements between the vegetation and soil. Journal of Ecology 70:649-666.

Elkhatib, E. A., Bennett, O. L., Baligar, and Wright, R. J. 1986. A centrifuge method for obtaining soil solution using an immiscible liquid. Soil Science Society of America Journal 50:297-299.

Elkins, N. Z, Sabol, G. V., Ward, T. J., and Whitford, W. G. 1986. The influence of subterranean termites on the hydrological characteristics of a Chihuahuan desert ecosystem. Oecologia (Berlin) 68:521-528.

Elliott, E. T. 1986. Aggregate structure and carbon, nitrogen, and phosphorus in native and cultivated soils. Soil Science Society of America Journal 50:627-633.

Elliott, E. T. and Coleman, D. C. 1988. Let the soil work for us. Ecological Bulletins (Copenhagen) 39:23-32.

Enwezor, W. O. and Moore, A. W. 1966. Phosphorus status of some Nigerian soils. Soil Science 102:322-328.

EPA. 1988. Stabilization report. Workshop on agriculture and climate change. U.S. Environmental Protection Agency, Washington, D.C. 45 pp.

Evans, L. J. and Wilson, W. G. 1985. Extractable Fe, Al, Si and C in B horizons of podzolic and brunisolic soils from Ontario. Canadian Journal of Soil Science 65:489-496.

FAO. 1985. Yearbook of agriculture for 1984. FAO, Rome.

———. 1978. Organic recycling in Asia. Soils Bulletin 36. FAO, Rome. 417 pp.

———. 1975. Organic materials as fertilizers. Soils Bulletin 27. FAO, Rome. 394 pp.

FAO-UNESCO. 1974. Soil map of the world, 1:5,000,000. Vol. 1. Legend. UNESCO, Paris.

van Fassen, H. G. and Smilde, K. W. 1985. Organic matter and turnover in soils. In: B. T. Kang and J. van der Heide (eds.), Nitrogen management in farming systems in humid and subhumid tropics. Institute for Soil Fertility, Haren, The Netherlands. pp. 39-55.

Flaig, W. 1975. Chemical composition and physical properties of humic substances. In: J. E. Gieseking (ed.), Soil components, vol 1. Springer-Verlag, New York.

Fogel, R. 1980. Mycorrhizae and nutrient cycling in natural forest ecosystems. New Phytologist 86:199-212.

Ford, G. W., Greenland, D. J., and Oades, J. M. 1969. Separation of the light fraction from soils by ultrasonic dispersion in halogenated hydrocarbons containing a surfactant. Journal of Soil Science 20:291-296.

Fox, R. H. 1980. Soils with variable charge: agronomic and fertility aspects. In: B. K. G. Theng (ed.), Soils with variable charge. New Zealand Society of Soil Science, Lower Hutt. pp. 195-224.

Fox, R. H. and Bandel, V. A. 1986. Nitrogen utilization with no-tillage. In: M. A. Sprague and G. B. Triplett (eds.), No-tillage and surface-tillage agriculture: the tillage revolution. Wiley, New York. pp. 117-148.

Fox, R. H. and Searle, P. G. E. 1978. Phosphate adsorption by soils of the tropics. In: M. Drosdoff (ed.), Diversity of soils in the tropics. Special Publication 34. American Society of Agronomy, Madison, Wisconsin. pp. 97-119.

Fragoso, C. and Lavelle, P. 1987. The earthworm community of a Mexican tropical rainforest (Chajul, Chiapas). In: A. M. Bonvieini and P. Omodeo (eds.), On earthworms. Selected Symposia and Monographs U.Z.I. 2, Mucchi, Model A. pp. 281-295.

Freney, J. R. 1986. Forms and reactions of organic sulfur compounds in soils. In: M. A. Tabatabai (ed.), Sulfur in agriculture. American Society of Agronomy, Madison, Wisconsin. pp. 207-232.

Frissel, M. J. (ed.) 1978. Cycling of mineral nutrients in agricultural ecosystems. Elsevier, Amsterdam. 356 pp.

Gartlan, J. S., Newberry, D. McC., Thomas, D. W., and Waterman, P. G. 1986. The influence of topography and soil phosphorus on the vegetation of Korup Forest Reserve, Cameroun. Vegetation 65:131-148.

Gillman, G. P. 1974. The influence of net charge on water dispersible clay and sorbed sulphate. Australian Journal of Soil Research 12:173-176.

Goedert, W. J. (ed.) 1986. Solos dos Cerrados: technologías e estrategias de manejo. Nobel, São Paulo. 422 pp.

Goh, K. M., Rafter, T. A., Stout, J. D., and Walker, T. W. 1976. The accumulation of soil organic matter and its carbon isotope content in a chronosequence of soils developed on aeolian sand in New Zealand. Journal of Soil Science 27:89-100.

Gonzalez, M. A. and Sauerbeck, D. R. 1982. Decomposition of ^{14}C labelled plant residues in different soils and climates of Costa Rica. Proceedings of the Regional Colloquium on Soil Organic Matter Studies, Piracicaba, SP-Brazil. PROMOCET, São Paulo. pp. 141-146. Available through Centro de Energia Nuclear na Agricultura, Av. Centenario, Caixa Postal 96, 13400 Piracicaba, S.P., Brazil.

Gould, E., Andau, M., and Easton, E. G. 1987. Observations of earthworms in Sepilok Forest, Sabah, Malaysia. Biotropica 19:370-372.

Gower, S. T. 1987. Relations between mineral nutrient availability and fine root biomass in two Costa Rican wet forests: a hypothesis. Biotropica 19:171-175.

Greenland, D. J. and Nye, P. H. 1959. Increases in carbon and nitrogen contents of tropical soils under natural fallows. Journal of Soil Science 10:284-299.

Griffith, E. and Burns, R. G. 1972. Interaction between phenolic substances and microbial polysaccharides in soil aggregation. Plant and Soil 36:599-612.

Grove, T. L. 1985. Phosphorus cycles in forests, grasslands, and agricultural ecosystems. Ph.D. thesis. Cornell University, Ithaca, New York.

Guar, A. C. 1984. Response of rice to organic matter: the Indian experience. In: Organic matter and rice. IRRI, Los Banos, The Philippines. pp. 503-514.

Gupta, S. R., Rajvanshi, R., and Singh, J. S. 1981. The role of the termite *Odontotermes gurdaspurensis* (Isoptera, Termitidae) in plant litter decomposition in a tropical grassland. Pedobiologia 22:254-261.

Haider, K. and Martin, J. P. 1981. Biochemistry of humus formation and its interaction with clays. In: Migrations organo-minérales dans les sols tempérés, Nancy, 1979. Centre National de la Recherche Scientifique, Paris. pp. 163-172.

Haines, B. L. 1978. Element and energy flows through colonies of the leaf-cutting ant, Atta colombica, in Panama. Biotropica 10:270-277.

Hairiah, K. and van Noordwijk, M. 1986. Root studies on a tropical Ultisol in relation to nitrogen management. Report 7-86. Institute for Soil Fertility, Haren, The Netherlands. 116 pp.

Halma, G., Posthumus, M. A., Miedman, R., van de Westeringh, R., and Meuzelaar, H. L. C. 1978. Characterisation of soil types by pyrolysis-mass spectrometry. Agrochimica 22:372-382.

Halstead, R. L. and McKercher, R. B. 1975. Biochemistry and cycling of phosphorus. In: E. A. Paul and A. D. McLaren (eds.), Soil Biochemistry, vol. 4. Dekker, New York.

Hamblin, A. P. and Greenland, D. J. 1977. Effects of organic constituents and complexed metal ions on aggregate stability of some East Anglian soils. Journal of Soil Science 28:410-416.

Handley, W. R. C. 1954. Mull and mor formation in relation to forest soils. Bulletin 23. Forestry Commission, H.M.S.O., Oxford, England. 115 pp.

Harmsen, G. W. 1951. Die Bedeutung der Bodenoberflache fur die Humusbildung. Plant and Soil 3:110. (as cited by Stevenson, 1986)

Hatcher, P. G. and Spiker, E. C. 1988. Selective degradation of plant biomolecules. In: F. H. Frimmel and R. F. Christman (eds.), Humic substances and their role in the environment. Dahlem Workshop Report. Wiley, New York. pp. 59-74.

Hatcher, P. G., Vanderhart, D. L., and Earl, W. L. 1980. Use of solid-state ^{13}C NMR in structural studies of humic acids and humin from Holocene sediments. Organic Geochemistry 2:87-92.

Hatcher, P. G., Breger, I. A., Maciel, G. E., and Szeverenyi, N. M. 1985. Geochemistry of humin. In: G. R. Aiken, D. M. McKnight, R. L. Wershaw, and P. MacCarthy (eds.), Humic substances in soil, sediment, and water. Wiley, New York. pp. 275-302.

Hatcher, P. G., Schnitzer, M., Dennis, L. W., and Maciel, G. E. 1981. Aromaticity of humic substances in soil. Soil Science Society of America Journal 45:1089-1094.

Hawkes, G. E., Powlson, D. S., Randall, E. W., and Tate, K. R. 1984. A ^{31}P nuclear magnetic resonance study of the phosphorus species in alkali extracts from long-term field experiments. Journal of Soil Science 35:35-45.

Hawkins, P. and Brunt, M. 1965. Report to the government of Cameroon on the soils and ecology of West Cameroon. Report 2083. FAO, Rome.

Hayes, M. H. B. 1985. Extraction of humic substances from soil. In: G. R. Aiken, D. M. McKnight, R. L. Wershaw, and P. MacCarthy (eds.), Humic substances in soil, sediment, and water. Wiley, New York. pp. 329-362.

Hayes, M. H. B. and Swift, R. S. 1978. The chemistry of soil organic colloids. In: D. J. Greenland and M. H. B. Hayes (eds.), The chemistry of soil constituents. Wiley, Chichester, England. pp. 179-320.

Hayes, M. J. and Cooley, J. H. 1987. Tropical Soil Biology: current status of concepts. Intecol Bulletin 14. University of Georgia, Athens, Georgia. 59 p.

He, X.-T., Stevenson, F. J., Mulvaney, R. L., and Kelley, K. R. 1988a. Incorporation of newly immobilized ^{15}N into stable organic forms in soil. Soil Biology and Biochemistry 20:74-81.

He, X.-T., Stevenson, F. J., Mulvaney, R. L., and Kelley, K. R. 1988b. Extraction of newly immobilized ^{15}N from an Illinois Mollisol using aqueous phenol. Soil Biology and Biochemistry 20:857-862.

Hedley, M. J. and Stewart, J. W. B. 1982. Method to measure microbial biomass phosphorus in soils. Soil Biology and Biochemistry 14:337-385.

Hedley, M. J., Stewart, J. W. B., and Chauhan, B. S. 1982. Changes in inorganic and organic soil phosphorus fractions induced by cultivation practices and by laboratory incubations. Soil Science Society of America Journal 46:970-976.

Helal, H. M. and Sauerbeck, D. R. 1986. Effect of plant roots on carbon metabolism of soil microbial biomass. Zeitschrift für Pflanzenernahrung und Bodenkunde 149:181-188.

Helling, C. S., Chesters, G., and Corey, R. B. 1964. Contribution of organic matter and clay to soil cation exchange capacity as affected by pH of the saturating solution. Soil Science Society of America Proceedings 28:517-520.

Hempfling, R., Ziegler, F., Zech, W., and Schulten, H. R. 1987. Litter decomposition and humification in acidic forest soils studied by chemical degradation, IR and NMR spectroscopy and pyrolysis field ionization mass spectrometry. Zeitschrift für Pflanzenernahrung und Bodenkunde 150:179-186.

Henderson, J. B. and Kamprath, E. J. 1970. Nutrient and dry matter accumulation by soybeans. Technical Bulletin 197. North Carolina Agricultural Experiment Station, Raleigh. 27 pp.

Hendrix, P. F., Parmelee, R. W., Crossley, D. A., Jr., Coleman, D. C., Odum, E. P., and Groffman, P. M. 1986. Detritus food webs in conventional and no-tillage agroecosystems. Bioscience 36:374-380.

Herbillon, A. J. 1980. Mineralogy of Oxisols and oxic materials. In: B. K. G. Theng (ed.), Soils with variable charge. New Zealand Society of Soil Science, Lower Hutt. pp. 109-126.

Higashi, T. 1983. Characterization of Al/Fe-humus complexes in Dystrandepts through comparison with synthetic forms. Geoderma 31:277-288.

Hinds, A. A. and Lowe, L. E. 1980. Distribution of carbon, nitrogen, sulfur, and phosphorus in particle size-separates from gleysolic soils. Canadian Journal of Soil Science 60:783-786.

Hoffman, G. and de Leenheer, L. 1975. Influence of soil pre-wetting on aggregate instability. Pedologie 25:190-198.

Holdridge, L. R. 1967. Life zone ecology. Tropical Science Center, San José, Costa Rica. 206 pp.

Holland, E. A. and Coleman, D. C. 1987. Litter placement effects on microbial and organic matter dynamics in an agroecosystem. Ecology 68:425-433.

Hopkins, C. B. 1910. Soil fertility and permanent agriculture. Ginn and Co., Boston. 651 pp.

Houghton, R. A. and Woodwell, G. M. 1989. Global climate change. Scientific American 260:36-44.

House, G. J., Stinner, B. J., Crossley, D. A. Jr., Odum, E. P., and Langdale, G. W. 1984. Nitrogen cycling in conventional and no-tillage agroecosystems in the Southern Piedmont. Journal of Soil and Water Conservation 39:194-200.

Howard-Williams, C. 1974. Nutritional quality and calorific value of Amazonian forest litter. Amazonia 1:67-75.

Huang, P. H. and Schnitzer, M. (eds.) 1986. Interactions of soil minerals with natural organics and microbes. Special Publication 17. Soil Science Society of America, Madison, Wisconsin. 606 pp.

Hue, N. V., Craddock, G., and Adams, F. 1986. Effects of organic acids on aluminum toxicity in subsoils. Soil Science Society of America Journal 50:28-34.

Hughes, J. C. 1982. High gradient magnetic separation of some soil clays from Nigeria, Brazil and Colombia. 1. The interrelationships of iron and aluminum extracted by acid ammonium oxalate and carbon. Journal of Soil Science 33:509-519.

Hunt, H. W., Stewart, J. W. B., and Cole, C. V. 1983. A conceptual model for interactions among carbon, nitrogen, sulphur and phosphorus in grasslands. In: B. Bolin and R. B. Cook (eds.), The major biogeochemical cycles and their interactions. Wiley, New York.

Ino, Y. and Monsi, M. 1969. An experimental approach to the calculation of CO_2 amount evolved from several soils. Japanese Journal of Botany 20:153-188.

Islam, A. and Ahmed, B. 1973. Distribution of inositol phosphates, phospholipids, and nucleic acids and mineralization of inositol phosphates in some Bangladesh soils. Journal of Soil Science 24:193-198.

Jager, G. 1971. The effect of living roots and the rhizosphere microflora on the decomposition of organic matter. In: Proceedings of the 4th International Congress of Soil Zoology. INRA, Paris. pp. 57-67.

Jansson, S. L. 1958. Tracer studies on nitrogen transformations in soil with special attention to mineralization-immobilization relationships. Annals of the Royal Agricultural College, Sweden 24:101-361.

Janzen, H. H. 1987. Soil organic matter characteristics after long-term cropping to various spring wheat rotations. Canadian Journal of Soil Science 67:845-856.

Jenkinson, D. S. 1971. Studies on the decomposition of ^{14}C labelled organic matter in soil. Soil Science 111:64-70.

Jenkinson, D. S. and Ayanaba, A. 1977. Decomposition of carbon-14 labeled plant material under tropical conditions. Soil Science Society of America Journal 41:912-915.

Jenkinson, D. S. and Ladd, J. N. 1981. Microbial biomass in soil: measurement and turnover. In: E. A. Paul and J. N. Ladd (eds.), Soil biochemistry, vol. 5. Dekker, New York. pp. 415-471.

Jenkinson, D. S. and Powlson, D. S. 1976. The effects of biocidal treatments on metabolism in soil. V. A method of measuring soil biomass. Soil Biology and Biochemistry 8:209-213.

Jenkinson, D. S. and Rayner, J. H. 1977. The turnover of soil organic matter in some of the Rothamsted classical experiments. Soil Science 123:298-305.

Jenkinson, D. S., Powlson, D. S., and Wedderburn, R. W. M. 1976. The effects of biocidal treatments on metabolism in soil. III. The relationship between soil biovolume, measured by optical microscopy, and the flush of decomposition caused by fumigation. Soil Biology and Biochemistry 8:189-202.

Jenkinson, D. S., Hart, P. B. S., Rayner, J. H., and Parry, L. C. 1987. Modelling the turnover of organic matter in long-term experiments at Rothamsted. IN-TECOL Bulletin 15:1-8.

Jenny, H. 1941. Factors of soil formation. McGraw-Hill, New York.

Jensen, H. L. 1929. On the influence of the carbon: nitrogen ratios of organic material on the mineralisation of nitrogen. Journal of Agricultural Science 19:71-82.

Johnston, A. E. and Mattingly, G. E. G. 1976. Experiments on the continuous growth of arable crops at Rothamsted and Woburn experimental stations: effects of treatments on crop yields and recent modifications in purpose and design. Annals of Agronomy 27:927-956.

Jones, J. A. 1988. Termites, soil fertility and C cycling in a dry African woodland: a hypothesis. Journal of Tropical Ecology. Updating.

Jones, J. A. 1973. The organic matter content of the savanna soils of West Africa. Journal of Soil Science 24:42-53.

Jones, M. J. 1971. The maintenance of soil organic matter under continuous cultivation at Sameru, Nigeria. Journal of Agricultural Science (Camb.) 77:473-482.

Jones, M. J. and Wild, A. 1975. Soils of the West African savanna. Commonwealth Agricultural Bureau, The Cambrian News, Aberystwyth, Wales. 246 pp.

Jordan, C. F. 1985. Nutrient cycling in tropical forest ecosystems. Wiley, Chichester, England.

Jordan, C. F. and Escalante, G. 1980. Root productivity in an Amazonian rainforest. Ecology 61:14-18.

Juma, N. G. and McGill, W. B. 1986. Decomposition and nutrient cycling in agroecosystems. In: M. J. Mitchell and J. P. Nakas (eds.), Microfloral and faunal interactions in natural and agroecosystems. Martinus Nijhoff, Dordrecht, The Hague, The Netherlands. pp. 74-136.

Juma, N. G., Paul, E. A., and Mary, B. 1984. Kinetic analysis of net nitrogen mineralization in soil. Soil Science Society of America Journal 48:753-757.

Juste, C., Delas, J., and Langon, M., 1975. Comparaison de la stabiliée biologique de différents humates metalliques. Comptes Rendus de l'Académie des Sciences, Paris, D. 281:1685-1688.

Kang, B. T. and van der Heide, J. (ed.) 1985. Nitrogen management in farming systems in humid and subhumid tropics. Institute for Soil Fertility, Haren, The Netherlands. 362 pp.

Kavanagh, B. V., Posner, A. M., and Quirk, J. P. 1976. The adsorption of polyvinyl alcohol on gibbsite and goethite. Journal of Soil Science 27:467-477.

Kelley, K. R. and Stevenson, F. J. 1987. Effects of carbon source on immobilization and chemical distribution of fertilizer nitrogen in soil. Soil Science Society of America Journal 51:946-951.

Kelley, K. R. and Stevenson, F. J. 1985. Characterization and extractability of immobilized ^{15}N from the soil microbial biomass. Soil Biology and Biochemistry 17:517-523.

Kemper, W. D. and Rosenau, R. C. 1986. Aggregate stability and size distribution. In: A. Klute (ed.), Methods of soil analysis. Part 1. 2nd ed. American Society of Agronomy, Madison, Wisconsin. pp. 425-442.

Keyes, M. R. and Grier, C. C. 1981. Above and below-ground net production in 40-year-old Douglas-fir stands on low and high productivity sites. Canadian Journal of Forest Research 11:599-605.

Khatibu, A. I., Lal, R., and Jana, R. K. 1984. Effects of tillage methods and mulching on erosion and physical properties of a sandy clay loam in an Equatorial warm humid region. Field Crops Research 8:239-254.

Kigne, J. W. 1968. Heats of wetting and complexes between montmorillonite and alkylammonium compounds. Transactions of the 9th International Congress of Soil Science (Adelaide, Australia) 1:597-605.

Kigne, J. W. and Taylor, S. A. 1964. Heat of wetting organo-montmorillonite complexes. Transactions of the 8th International Congress of Soil Science (Bucharest, Romania) 2:205-209.

Kitazawa, Y., 1967. Community metabolism of soil invertebrates in forest ecosystems of Japan. In: K. Petrusewicz (ed.), Secondary productivity of terrestrial ecosystems. Polish Academy of Science, Warsaw. pp. 649-661.

Kogel, I., Hempfling, R., Zech, W., Hatcher, P. G., and Schulten, H-R. 1988. Chemical composition of the organic matter in forest soils. 1. Forest litter. Soil Science 146:124-136.

Kononova, M. M. 1975. Humus of virgin and cultivated soils. In: J. E. Gieseking (ed.), Soil components. Vol. 1. Organic components. Springer-Verlag, New York. pp. 475-525.

————. 1966. Soil organic matter. 2nd English ed. Pergamon, London.

Kucey, R. M. N. and Paul, E. A. 1982. Carbon flow, photosynthesis and N_2 fixation in mycorrhizal and nodulated faba beans (*Vicia faba* L.). Soil Biology and Biochemistry 14:407-412.

Kumada, K., 1987. Chemistry of soil organic matter. Developments in soil Science 17. Japan Scientific Societies Press, Tokyo/Elsevier, Amsterdam. 241 pp.

Kumazawa, K. 1984. Beneficial effects of organic matter on rice growth and yield in Japan. In: Organic matter and rice. IRRI, Los Banos, The Philippines. pp. 431-454.

Kummert, R. and Stumm, W. 1980. The surface complexation of organic acids on hydrous δ-Al_2O_3. Journal of Colloid Interface Science 75:373-385.

Ladd, J. N., Oades, J. M., and Amato, M. 1981. Microbial biomass formed from ^{14}C-, ^{15}N-labelled plant material decomposing in soils in the field. Soil Biology and Biochemistry 13:119-126.

Ladd, J. N., Parsons, J. W., and Amato, M. 1977a. Studies of nitrogen immobilization and mineralization in calcareous soils. I. Distribution of immobilized nitrogen among soil fractions of different particle size and density. Soil Biology and Biochemistry 9:309-318.

Ladd, J. N., Parsons, J. W., and Amato, M. 1977b. Studies of nitrogen immobilization and mineralization in calcareous soils. II. Mineralization of immobilized nitrogen from soil fractions of different particle size and density. Soil Biology and Biochemistry 9:319-326.

Ladd, J. N., Jackson, R. B., Amato, M., and Butler, H. H. A. 1983. Decomposition of plant material in Australian soils. I. The effect of quantity added on decomposition and on residual microbial biomass. Australian Journal of Soil Research 21:563-570.

Laishram, I. D. and Yadava, P. S. 1988. Lignin and nitrogen in the decomposition of leaf litter in a subtropical forest ecosystem at Shiroy hills in north-eastern India. Plant and Soil 106:59-64.

Lal, R. 1987. Tropical ecology and physical edaphology. Wiley, New York.

————. 1984. Soil erosion from tropical arable lands and its control. Advances in Agronomy 37:183-248.

————. (ed.) 1979. Soil tillage and crop production. Proceedings Series 2. IITA, Ibadan, Nigeria.

————. 1976. Soil erosion problems on an Alfisol in Western Nigeria and their control. Monograph 1. IITA, Ibadan, Nigeria. 208 pp.

Lal, R. and Kang, B. T. 1982. Management of organic matter in soils of the tropics and subtropics. Transactions of the 12th International Congress of Soil Science, New Dehli, India. 4:152-178.

Lal, R., De Vleeschauwer, D., and Nganje, R. M. 1980. Changes in properties at a newly cleared Alfisol as affected by mulching. Soil Science Society of America Journal 44:827-833.

Larson, W. E., Clapp, C. E., Pierre, W. H., and Morachan, Y. B. 1972. Effects of increasing amounts of organic residues on continuous corn. II. Organic carbon, nitrogen, phosphorus, and sulfur. Agronomy Journal 64:204-208.

Larson, W. P., Holt, R. F., and Carlson, C. W. 1978. Residues for soil conservation. Special Publication 31. American Society of Agronomy, Madison, Wisconsin. pp. 1-16.

Lathwell, D. J. and Bouldin, D. R. 1981. Soil organic matter and nitrogen behavior in cropped soils. Tropical Agriculture (Trinidad) 58:341-348.

Lavelle, P. 1988. Earthworm activities and the soil system. Biology and Fertility of Soils 6:237-251.

————. 1987. Biological processes and productivity of soils in the humid tropics. In: R. E. Dickinson (ed.), Geophysiology of Amazonia. Wiley, Chichester, England. pp. 175-223.

————. 1984. The soil system in the humid tropics. Biology International 9:2-20.

————. 1978. Les vers de terre de la savane de Lamto (Cote d'Ivoire): peuplements, populations et fonctions dans l'écosystème. Publication 12. Laboratoire de Zoologie, ENS, Paris. 301 pp.

Lavelle, P., Zaidi, Z., and Schaefer, R. 1983. Interactions between earthworms, soil organic matter and microflora in an African savanna soil. In: P. Lebrun, H. M. Andre, A. de Medts, C. Gregoire-Wibo, and G. Wauthy (eds.), New trends in soil biology. Dieu-Brichart, Louvain-la-Neuve, Belgium. pp. 254-259.

Lawes, J. B. 1889. A history of a field newly laid down to permanent grass. Journal of the Royal Agricultural Society of England. 2nd series 25:1-24.

Lawson, T. L. and Lal, R. 1979. Response of maize to surface and buried straw mulch on a tropical Alfisol. In: R. Lal (ed.), Soil tillage and crop production. Proceedings Series 2. IITA, Ibadan, Nigeria. pp. 63-74.

Lee, K. E. 1985. Earthworms: their ecology and relationships with soils and land use. Academic Press, London and New York.

————. 1983. Soil animals and pedological processes. In: Soils: an Australian viewpoint. CSIRO, Melbourne/Academic Press, London. pp. 629-644.

Lee, K. E. and Wood, T. G. 1971. Termites and soils. Academic Press, London. 251 pp.

Legay, B. and Schaefer, R. 1984. Modalities of the energy flow in different tropical soils, as related to their mineralization capacity of organic carbon and to the type of clay. Zentralblatt für Mikrobiologie 139:389-400.

Legay, B. and Schaefer, R. 1981. Modalities of energy-flow in different tropical soils, as related to their mineralization capacity of organic C and to the type of clay. I. Effect of water content and the role of thermic conditions. Zentralblatt für Mikrobiologie 136:461-470.

Levin, D. A. 1971. Plant phenolics: an ecological perspective. American Naturalist 105:157-181.

van der Linden, M. J. H. A. 1971. Availability of protein in leaf litter; an enzymalogical approach. In: Proceedings of the 4th International Congress of Soil Zoology. INRA, Paris. pp. 337-348.

Lindstrom, M. J. and Holt, R. F. 1983. Crop residue removal: the effect on soil erosion and nutrient loss. In: R. R. Lowrance, R. L. Todd, L. E. Asmussen, and R. A. Leonard (eds.), Nutrient cycling in agricultural ecosystems. Special Publication 23. University of Georgia Agricultural Experiment Station, Athens. pp. 427-439.

Lobartini, J. C. and Tan, K. H. 1988. Differences in humic acid characteristics as determined by carbon-13 nuclear magnetic resonance, scanning electron microscopy, and infrared analysis. Soil Science Society of America Journal 52:125-130.

Loomis, R. S. 1978. Ecological dimensions of medieval agrarian systems: an ecologist responds. Agricultural History 52(3):478-483.

Lopes, A. S. 1983. Solos sob Cerrados. Instituto da Potassa e Fosfato, Piracicaba, Brazil. 162 pp.

Lopes, A. S. and Cox, F. R. 1977. A survey of the fertility status of surface soils under "Cerrados" vegetation in Brazil. Soil Science Society of America Journal 41:742-747.

Lowe, L. E. and Hinds, A. A. 1983. The mineralization of nitrogen and sulfur from particle-size separates of gleysolic soils. Canadian Journal of Soil Science 63:761-766.

Luxton, M. 1982. General ecological influence of the soil fauna on decomposition and nutrient circulation. Oikos 39:355-388.

Luxton, M. and Petersen, H. 1982. Survey of the main animal taxa of the detritus food web. Oikos 39:293-294.

Lynch, J. M. and Bragg, E. 1985. Microorganisms and soil aggregate stability. In: B. A. Stewart (ed.), Advances in soil science, vol. 2. Springer-Verlag, New York.

Lynch, J. M. and Panting, L. M. 1980. Cultivation and the soil biomass. Soil Biology and Biochemistry 12:29-33.

Mahareswaran, J. and Gunatilleke, I. U. N. 1988. Litter decomposition in a lowland rainforest and a deforested area in Sri Lanka. Biotropica 20(1):90-99.

Mariotti, A. and Guillet, B. 1987. Natural carbon-13 abundance as a tracer for studies of soil organic matter dynamics. Soil Biology and Biochemistry 19:25-30.

Marsh, K. B., Tillman, R. W., and Syers, J. K. 1987. Charge relationships of sulfate sorption by soils. Soil Science Society of America Journal 51:318-323.

Marshall, J. D. and Waring, R. H. 1985. Predicting fine root production and turnover by monitoring root starch and soil temperature. Canadian Journal of Forest Research 15:791-800.

Martens, R. 1985. Limitations in the application of the fumigation technique for biomass estimation in amended soils. Soil Biology and Biochemistry 17:57-63.

Martin, A., Cortez, J., Barois, I., and Lavelle, P. 1987. Les mucus intestinaux de vers de terre moteurs de leure interactions avec la microflore. Rev. Ecol. Biol. Sol. 24(3):549-558.

Martin, J. K. 1977. Factors influencing the loss of organic carbon from wheat roots. Soil Biology and Biochemistry 9:1-17.

Martin, J. P. and Haider, K. 1980. Microbial degradation and stabilization of ^{14}C-labelled lignins, phenols, and phenolic polymers in relation to soil humus formation. In: T. K. Kirk, T. Higuchi, and H. M. Chang (eds.), Lignin biodegradation: microbiology chemistry and potential applications, vol. 2. CRC Press, West Palm Beach, Florida. pp. 77-100.

Martin, J. P. and Haider, K. 1977. Decomposition in soil of specifically ^{14}C-labelled DHP and corn stalk lignins, model humic acid-type polymers and coniferyl alcohols. In: Soil organic matter studies. Proceedings of the FAO/IAEA Symposium, Braunschweig. IAEA, Vienna. 2:23-32.

Martin, J. P., Ervin, J. O., and Shepherd, R.A. 1966. Decomposition of the iron, aluminum, zinc, and copper salts or complexes of some microbial and plant polysaccharides in soil. Soil Science Society of America Proceedings 30:196-200.

Martin, J. P., Zunino, H., Peirano, P., Caiozii, M., and Haider, K. 1982. Decomposition of ^{14}C-labelled lignins, model humic acid polymers, and fungal melanins in allophanic soils. Soil Biology and Biochemistry 14:289-293.

Martin, J. S. and Martin, M. M. 1982. Tannin assays in ecological studies: lack of correlation between phenolics, proanthocyanidins and protein-precipitating constituents in mature foliage of six oak species. Journal of Chemical Ecology 9:285-294.

Martinez-Ramos, M., Alvarez-Buylla, E., Sarukhan, J., and Pinero, D. 1988. Treefall age determination and gap dynamics in a tropical forest. Journal of Ecology 76:700-716.

Matsumoto, T. and Abe, T. 1979. The role of termites in an equatorial rain forest ecosystem in West Malaysis. II. Leaf litter consumption on the forest floor. Oecologia (Berlin) 38:261-274.

Maurya, P. R. and Lal, R. 1981. Effects of different mulch materials on soil properties and on the root growth and yield of maize and cowpea. Field Crops Research 4:33-45.

McClaugherty, C. A., Aber, J. D., and Melillo, J. M. 1982. The role of fine roots in the organic matter and nitrogen budgets of two forested ecosystems. Ecology 63:1481-1490.

McGill, W. B. and Cole, C. V. 1981. Comparative aspects of organic C, N, S and P cycling through organic matter during pedogenesis. Geoderma 26:267-286.

McGill, W. B. and Myers, R. J. K. 1987. Controls on dynamics of soil and fertilizer nitrogen. In: R. F. Follett, J. W. B. Stewart, and C. V. Cole (eds.), Soil fertility and organic matter as critical components of production systems. Special Publication 19. Soil Science Society of America, Madison, Wisconsin. pp. 73-79.

McGill, W. B. and Paul, E. A. 1976. Fractionation of soil and ^{15}N nitrogen to separate the organic and clay interactions of immobilized N. Canadian Journal of Science 56:203-212.

McGill, W. B., Cannon, K. R., Robertson, J. A., and Cook, F. D. 1986. Dynamics of soil microbial biomass and water-soluble organic C in Breton L after 50 years of cropping to two rotations. Canadian Journal of Soil Science 66:1-19.

McGill, W. B., Hunt, H. W., Woodmansee, R. G., and Reuss, J. O. 1981. PHOENIX, a model of the dynamics of carbon and nitrogen in grassland soils. In: F. E. Clark and T. Rosswall (eds.), Terrestrial nitrogen cycles. Ecological Bulletins (Stockholm) 33:49-115.

McKercher, R. B. and Anderson, G. 1968. Content of inositol penta- and hexa-phosphates in some Canadian soils. Journal of Soil Science 19:47-55.

McLaren, R. G. and Crawford, D. V. 1973. Studies on soil copper. 1. The fractionation of copper in soil. Journal of Soil Science 24:172-181.

McLaughlin, M. J., Alston, A. M., and Martin, J. K. 1988. Phosphorus cycling in wheat-pasture rotations. III. Organic phosphorus turnover and phosphorus cycling. Australian Journal of Soil Research 26:343-353.

McNaughton, S. J. 1983. Serengeti grassland ecology: the role of composite environmental factors and contingency in community organization. Ecological Monographs 53:291-320.

McVey, D., Watermann, P. G., Mbi, C. N., Gartlan, J. S., and Struhsaker, T. T. 1978. Phenolic content of vegetation in two African rain forests: ecological interpretations. Science 202:61-64.

Meentemeyer, V. 1978. Macroclimate and lignin control of decomposition rates. Ecology 59:465-472.

Melillo, J. M. and Gosz, J. R. 1983. Interactions of biogeochemical cycles in forest ecosystems. In: B. Bolin and R. B. Cook (eds.), The major biogeochemical cycles and their interactions. SCOPE Vol. 21. Wiley, Chichester, England. pp. 177-222.

Melillo, J. M., Aber, J. D., and Muratore, J. F. 1982. Nitrogen and lignin control of hardwood leaf litter decomposition dynamics. Ecology 63:621-626.

Milchunas, D. G., Lauenroth, W. K., Singh, J. S., Cole, C. V., and Hunt, H. W. 1985. Root turnover and production by ^{14}C dilution: implications of carbon partitioning in plants. Plant and Soil 88:353-365.

Mizota, C. and Chapelle, J. 1988. Characterization of some Andepts and andic soils in Rwanda, Central Africa. Geoderma 41:193-209.

Mizota, C., Carrasco, M. A., and Wada, K. 1982. Clay mineralogy and some chemical properties of Ap horizons of Ando soils used for paddy rice in Japan. Geoderma 27:225-237.

Mohr, E. C. J. 1930. Tropical soil farming processes and the development of tropical soils. National Geologic Survey of China, Beiping.

Mohr, E. C. J. and van Baren, F. A. 1954. Tropical soils. Van Hoeve, The Hague, The Netherlands.

Molina, J. A. E., Clapp, C. E., Schaffer, M. J., Chichester, F. W., and Larson, W. E. 1983. NCSOIL, a model of nitrogen and carbon transformations in soil; description, calibration and behavior. Soil Science Society of America Journal 47:85-91.

Molloy, L. F. and Blakemore, L. C. 1974. Studies on a climosequence of soils in tussock grasslands. 1. Introduction, sites, and soils. New Zealand Journal of Science 17:233-255.

Molloy, L. F. and Speir, T. W. 1977. Studies on a climosequence of soils in tussock grasslands. 12. Constituents of the soil light fraction. New Zealand Journal of Science 20:167-177.

Molloy, L. F., Bridger, B. A., and Cairns, A. 1977. Studies on a climosequence of soils in tussock grasslands. 13. Structural carbohydrates in tussock leaves, roots, and litter and in the soil light and heavy fractions. New Zealand Journal of Science 20:443-451.

Monroe, C. D. and Kladivko, E. J. 1987. Aggregate stability of a silt loam soil as affected by roots of corn, soybeans and heat. Communications in Soil Science and Plant Analysis 18:1077-1087.

Moshi, A. O., Wild, A., and Greenland, D. J. 1974. Effect of organic matter on the charge and phosphate adsorption characteristics of Kikuyu red clay from Kenya. Geoderma 22:275-285.

Mueller-Harvey, I., Juo, A. S. R., and Wild, A. 1985. Soil organic C, N, S and P after forest clearance in Nigeria: mineralization rates and spatial variability. Journal of Soil Science 36:585-591.

Muller, R. N., Kalisz, P. J., and Kimmerer, T. W. 1987. Intraspecific variation in production of astringent phenolics over a vegetation-resource availability gradient. Oecologia (Berlin) 72: 211-215.

Munevar, F. and Wollum, A. G. II. 1977. Effects of the additions of phosphorus and inorganic nitrogen on carbon and nitrogen mineralization in Andepts from Colombia. Soil Science Society of America Journal 41:540-545.

Murayama, S. 1988. Microbial synthesis of saccharides in soils incubated with ^{13}C-labelled glucose. Soil Biology and Biochemistry 20:193-199.

Nadelhoffer, K. J., Aber, J. D., and Melillo, J. M. 1985. Fine roots, net primary production, and soil nitrogen availability: a new hypothesis. Ecology 66:1377-1390.

Naske, A. and Richter, J. 1981. N mineralization in Loss-Parabrownearths: Incubation experiments. Plant and Soil 59:237-247.

Nemeth, A. 1981. Estudio ecologico de las lombrices de tierra (Oligochaeta) en ecosistemas de bosque humedo tropical en San Carlos de Rio Negro, Territorio Federal Amazonas. Tesis de Licenciatura, University of Central Venezuela. 92 pp.

Neptune, A. M. L., Tabatabai, M. A., and Hanway, J. J. 1975. Sulfur fractions and carbon-nitrogen-phosphorus-sulfur relationships in some Brazilian and Iowa soils. Soil Science Society of America Proceedings 39:51-54.

Newberry, D. McC. and Proctor, J. 1984. Ecological studies in four contrasting lowland rainforests in Gunung Mulu National Park, Sarawak. IV. Associations between tree distribution and soil factors. Journal of Ecology 72:475-493.

Newman, R. H., Theng, B. K. G., and Filip, Z. 1987. Carbon-13 nuclear magnetic resonance spectroscopic characterisation of humic substances from municipal refuse decomposing in a landfill. Science of the Total Environment 65:69-84.

Nip, M., Tegelaar, E. W., Brinkhuis, H., De Leeuw, J. W., Schenck, P. A., and Holloway, P. J. 1986. Analysis of modern and fossil plant cuticles by Curie point

Py-GC and Curie point Py-GC-MS: recognition of a new, highly aliphatic and resistant biopolymer. Organic Geochemistry 10:769-778.

van Noordwijk, M. 1987. Methods for quantification of root distribution pattern and root dynamics in the field. International Potash Institute, Baden bei Wien, Austria. pp. 243-262.

Nye, P. H. and Bertheux, M. H. 1957. The distribution of phosphorus in forest and savanna soils of the Gold Coast and its agricultural significance. Journal of Agricultural Science 49:141-159.

Nye, P. H. and Greenland, D. J. 1960. The soil under shifting cultivation. Technical Communication 51. Commonwealth Bureau of Soils, Slough, England.

Oades, J. M. 1988a. An introduction to organic matter in mineral soils. Chapter 3, in: S. B. Weed and J. B. Dixon (eds.), Minerals in soil environments, 2nd ed. Soil Science Society of America, Madison, Wisconsin. pp. 187-259.

————. 1988b. The retention of organic matter in soils. Biogeochemistry 5:35-70.

————. 1984. Soil organic matter and structural stability: mechanisms and implications for management. Plant and Soil 76:319-337.

————. 1972. Studies on soil polysaccharides. III. Composition of polysaccharides in some Australian soils. Australian Journal of Soil Research 10:113-126.

————. 1967. Carbohydrates in some Australian soils. Australian Journal of Soil Research 5:103-115.

Oades, J. M. and Ladd, J. N. 1977. Biochemical properties: carbon and nitrogen metabolism. In: J. S. Russell and E. S. Greachen (eds.), Soil factors in crop production in a semi-arid environment. University of Queensland Press, St. Lucia, Australia. pp. 127-160.

Oades, J. M. and Tipping, E. 1988. Interactions of humic substances with oxides. In: M. H. B. Hayes, P. McCarthy, R. L. Malcolm, and R. S. Swift (eds.), Humic substances in soil, sediment and water. III. Interactions with metals, minerals and organic chemicals. Proceedings 2nd Conference of the International Humic Substances Society, Birmingham, U.K., July 1984. Wiley, Chichester, England. In press.

Oades, J. M. and Turchenek, L. W. 1978. Accretion of organic carbon, nitrogen and phosphorus in sand and silt fractions of a red-brown earth under pasture. Australian Journal of Soil Research 16:351-354.

Oades, J. M., Vassallo, A. M., Waters, A. G., and Wilson, M. A. 1987. Characterization of organic matter in particle size and density fractions from a red-brown earth by solid-state ^{13}C N. M. R. Australian Journal of Soil Research 25:71-82.

Oades, J. M., Waters, A. G., Vassallo, A. M., Wilson, M. A., and Jones, G. P. 1988. Influence of management on the composition of organic matter in a red-brown earth as shown by ^{13}C nuclear magnetic resonance. Australian Journal of Soil Research 26:289-299.

Ochwald, W. R. (ed.) 1978. Crop residue management systems. Special Publication 31. American Society of Agronomy, Madison, Wisconsin. 248 pp.

Ogner, G. 1985. A comparison of four different raw humus types in Norway using chemical degradations and CPMAS ^{13}C NMR spectroscopy. Geoderma 35:343-353.

Oh, W. K. 1984. Effects of organic matter on rice production. In: Organic matter and rice. IRRI, Los Banos, The Philippines. pp. 477-488.

Ohiagu, C. E. and Wood, T. G. 1979. Grass production and decomposition in southern Guinea savanna, Nigeria. Oecologia (Berlin) 40.

Olsen, S. R. 1986. The role of organic matter and ammonium in producing high corn yields. In: Y. Chen and Y. Avnimelech (eds.), The role of organic matter in modern agriculture. Martinus Nijhoff, Dordrecht, The Netherlands. pp. 29-54.

Olson, R. A. and Englestad, O. P. 1972. Soil phosphorus and sulfur. In: Soils of the humid tropics. National Academy of Science, Washington, D.C. pp. 82-101.

Orem, W. H. and Hatcher, P. G. 1987. Solid-state ^{13}C NMR studies of dissolved organic matter in pore waters from different depositional environments. Organic Geochemistry 11:73-82.

Orlov, D. S. 1985. Humus acids of soils. [Translation from Russian.] Amerind Publishing Co. Pvt. Ltd., New Delhi, India.

Page, A. L., Miller, R. H., and Keeney, D. R. 1986. Methods of soil analysis: Part 2. 2nd ed. American Society of Agronomy. Madison, Wisconsin.

Palm, C. A. 1988. Mulch quality and nitrogen dynamics in an alley cropping system in the Peruvian Amazon. Ph.D. dissertation, Soil Science Department, North Carolina State University, Raleigh. 112 pp.

Pang, P. C. and Paul, E. A. 1980. Effects of vesicular-arbuscular mycorrhizae on ^{14}C and ^{15}N distribution in nodulated faba beans. Canadian Journal of Soil Science 60:241-250.

Pardo, M. T. and Guadalix, M. E. 1988. Surface charge characteristics of volcanic ash soils from Spain. Effect of organic matter and mineralogical composition. Communications in Soil Science and Plant Analysis 19:259-269.

Parfitt, R. L., Farmer, V. C., and Russell, J. D. 1977a. Adsorption of hydrous oxides. I. Oxalate and benzoate on goethite. Journal of Soil Science 28:29-39.

Parfitt, R. L., Fraser, A. R., Russell, J. D., and Farmer, V. C. 1977b. Adsorption on hydrous oxides. II. Oxalate benzoate and phosphate on gibbsite. Journal of Soil Science 28:40-47.

Parfitt, R. L., Fraser, A. R., and Farmer, V. C. 1977c. Adsorption on hydrous oxides. III. Fulvic acid and humic acid on goethite, gibbsite and imogolite. Journal of Soil Science 28:289-296.

Parker, J. C., Zelazny, L. W., Sampath, S., and Harris, W. G. 1979. A critical evaluation of the extension of zero point of charge (ZPC) theory to soil systems. Soil Science Society of America Journal 43:668-674.

Parr, J. F. and Papendick, R. I. 1978. Factors affecting the decomposition of crop residues by microorganisms. Special Publication 31. American Society of Agronomy, Madison, Wisconsin. pp. 101-1030.

Parsons, J. W. 1988. Isolation of humic substances from soils and sediments. In: F. L. Frimmel and R. F. Christman (eds.), Humic substances and their role in the environment. Wiley, New York. pp. 2-14.

Parton, W. J., Stewart, J. W. B., and Cole, C. V. 1988. Dynamics of C, N, P and S in grassland soils: a model. Biogeochemistry 5:109-131.

Parton, W. J., Anderson, D. W., Cole, C. V., and Stewart, J. W. B. 1983. Simulation of soil organic matter formations and mineralization in semiarid agroecosystems. In: R. R. Lowrance, R. L. Todd, L. E. Asmussen, and R. A. Leonard (eds.), Nutrient cycling in agricultural ecosystems. Special Publication 23. University of Georgia Agricultural Experiment Station, Athens. pp. 533-550.

Parton, W. J., Schimel, D. S., Cole, C. V., and Ojima, D. S. 1987. Analysis of factors controlling soil organic matter levels in Great Plains grasslands. Soil Science Society of America Journal 51:1173-1179.

Parton, W. J., Cole, C. V., Stewart, J. W. B., Ojima, D. S., and Schimel, D. S. 1989. Simulating regional patterns of soil C, N, and P dynamics in the U.S. central grasslands region. Plant and Soil. In Press.

Pashanasi, B. and Lavelle, P. 1987. Soil macrofauna as affected by management practices. In: TropSoils/NCSU technical report for 1986-1987. Tropical Soils Research Program, Department of Soil Science, North Carolina State University, Raleigh.

Pastor, J. and Post, W. M. 1986. Influence of climate, soil moisture and succession on forest carbon and nitrogen cycles. Biogeochemistry 2:3-27.

Pate, J. S., Layzell, D. B., and Atkins, C. A. 1979. Economy of carbon and nitrogen in a nodulated and non nodulated (NO_3 grown) legume. Plant Physiology 64:1083-1088.

Paul, E. A. 1984. Dynamics of organic matter in soils. Plant and Soil 76:275-285.

Paul, E. A. and Juma, N. G. 1981. Mineralization and immobilization of soil nitrogen by microorganisms. In: F. E. Clark and T. Rosswall (eds.), Terrestrial nitrogen cycles. Ecological Bulletin (Stockholm) 33:179-195.

Paul, E. A. and van Veen, J. A. 1979. The use of tracers to determine the dynamic nature of organic matter. In: J. R. Gasser (ed.), Modelling nitrogen from farm wastes. Applied Science Publishers, London. pp. 75-132.

Paul, E. A. and van Veen, J. A. 1978. The use of tracers to determine the dynamic nature of organic matter. Transactions of the 11th International Congress of Soil Science (Edmonton). 3:61-102.

Paul, E. A. and Voroney, R. P. 1983. Field interpretation of microbial biomass activity measurements. In: M. J. Klug and C. A. Reddy (eds.), Current perspectives in microbial ecology. American Society for Microbiology, Michigan State University, East Lansing.

Paustian, K. and Bonde, T. A. 1987. Interpreting incubation data on nitrogen mineralization from soil organic matter. INTECOL Bulletin 15:101-112.

Perfect, T. J., Cook, A. C., and Swift, M. J. 1980. The effects of changing agricultural practice on the biology of a forest soil in the sub-humid tropics. I. The soil fauna. In: J. I. Furtado (ed.), Tropical ecology and development. International Society of Tropical Ecology, Kuala Lumpur, Malaysia. pp. 531-540.

Perrott, K. W. and Sarathchandra, S. U. 1987. Nutrient and organic matter levels in a range of New Zealand soils under established pasture. New Zealand Journal of Agricultural Research 30:249-259.

Persson, H. 1983. The distribution and productivity of fine roots in boreal forests. Plant and Soil 71:87-101.

———. 1980. Spatial distribution of fine roots growth, mortality and decomposition in a young Scots pine stand in central Sweden. Oikos 34:77-80.

Petersen, H. 1982. The total soil fauna biomass and its composition. Oikos 39:330-339.

Phillipson, J. 1973. The biological efficiency of protein production by grazing and other land-based systems. In: J. G. W. Jones (ed.), The biological efficiency of protein production. Cambridge University Press, Cambridge, England. pp. 217-235.

Pomeroy, D. E. 1983. Some effects of mound-building termites on the soils of a semi-arid area of Kenya. Journal of Soil Science 34:555-570.

Post, W. M., Emanuel, W. R., Zinke, P. J., and Stangenberger, A. G. 1982. Soil carbon pools and world life zones. Nature 298:156-159.

Post, W. M., Pastor, J., Zinke, P. J., and Stangenberger, A. G. 1985. Global patterns of soil nitrogen storage. Nature 317:613-616.

Preston, C. M. and Ripmeester, J. A. 1983. ^{13}C labelling for NMR studies of soils: direct CPMAS NMR observation of ^{13}C-acetate transformation in a mineral soil. Canadian Journal of Soil Science 63:495-500.

Preston, C. M. and Ripmeester, J. A. 1982. Application of solution and solid-state ^{13}C NMR to four organic soils, their humic acids, fulvic acids, humins and hydrolysis residues. Canadian Journal of Spectroscopy 27:99-105.

Preston, C. M., Schnitzer, M., and Ripmeester, J. A. 1989. Carbon-13 CPMAS NMR and chemical investigations of HCL/HF deashing of humin from a mineral soil. Soil Science Society of America Journal. In Press.

Preston, C. M., Dudley, R. L., Fyfe, C. A., and Mathur, S. P. 1984. Effects of variations in contact times and copper contents on a ^{13}C CPMAS NMR study of samples of four organic soils. Geoderma 33: 245-253.

Preston, C. M., Shipitalo, S.-E., Dudley, R. L., Fyfe, C. A., Mathur, S. P., and Levesque, M. 1987. Comparison of ^{13}C CPMAS NMR and chemical techniques for measuring the degree of decomposition in virgin and cultivated peat profiles. Canadian Journal of Soil Science 67:187-198.

Proctor, J. 1983. Tropical forest litter. In: S. L. Sutton, T. C. Whitmore, and A. C. Chardwick (eds.), Tropical rain forest: ecology and management. Blackwell Scientific, Oxford, England, pp 263-273.

Proctor, J., Anderson, J. M., Chai, and Vallack, H. W. 1983. Ecological studies in four contrasting lowland rainforests in Gunung Mulu National Park, Sarawak. I. Forest environment, structure and floristics. Journal of Ecology 71:237-260.

Radcliffe, D. E. and Gillman, G. P. 1985. Surface charge characteristics of volcanic ash soils from the Southern Highlands of Papua, New Guinea. In: E. Fernandez Caldas and D. H. Yaalon (eds.), Volcanic soils. Catena supplement 7. Catena-Verlag.

Ramsay, A. J., Stannard, R. E., and Churchman, G. J. 1986. Effect of conversion from ryegrass pasture to wheat cropping on aggregation and bacterial populations in a silt loam soil in New Zealand. Australian Journal of Research 24:253-264.

Rice, C. W., Smith, M. S., and Blevins, R. L. 1986. Soil nitrogen availability after long-term continuous no-tillage and conventional corn production. Soil Science Society of America Journal 50:1206-1210.

Robertson, G. P. 1984. Nitrification and nitrogen mineralization in a lowland rainforest succession in Costa Rica, Central America. Oecologia (Berlin) 61:91-95.

Robertson, G. P., Herrera, R., and Rosswall, T. (eds.) 1982. Nitrogen cycling in ecosystems of Latin America and the Caribbean. Martinus Nijhoff, Dordrecht, The Netherlands. 430 pp.

Robertson, G. P., K. Schnurer, J., Clarholm, M., Bonde, T. A., and Rosswall, T. 1988. Microbial biomass in relation to C and N mineralization during laboratory incubations. Soil Biology and Biochemistry 20:281-286.

Ross, D. J. and Tate, K. R. 1982. Microbial biomass of some yellow-brown loams under pasture and maize cultivation. In: Soil groups of New Zealand, part 6. Yellow-brown loams. New Zealand Society of Soil Science, Lower Hutt. pp. 78-80.

Ross, D. J., Tate, K. R., Cairns, A., and Pansier, E. A. 1980. Microbial biomass estimations in soils from tussock grasslands by three biochemical procedures. Soil Biology and Biochemistry 12:375-383.

Rosswall, T. (ed.) 1980. Nitrogen cycling in West African ecosystems. Royal Swedish Academy of Sciences, Stockholm. 449 pp.

Rusek, J. 1986. Soil microstructures — contributions on specific soil organisms. In: Faunal influences on soil structure. Quaestiones Entomologicae 21:497-514.

Russell, E. W. 1981. Soil conditions and plant growth, 10th ed. Longman, New York.

Ryszkowski, L. 1985. Impoverishment of soil fauna due to agriculture. IN-TECOL Bulletin 12:7-17.

Safford, L. O. 1974. Effects of fertilization on fine root nutrients and biomass. Plant and Soil 40:349-363.

Saggar, S., Bettany, J. R., and Stewart, J. W. B. 1981. Measurement of microbial sulphur in soil. Soil Biology and Biochemistry 13:493-498.

Saiz-Jimenez, C. and de Leeuw, J. W. 1987. Chemical structure of soil humic acid as revealed by analytical pyrolysis. Journal of Analytical Pyrolysis 11:367-376.

Saiz-Jimenez, C. and de Leeuw, J. W. 1986a. Chemical characterization of soil organic matter fractions by analytical pyrolysis-gas chromatography-mass spectrometry. Journal of Analytical and Applied Pyrolysis 9:99-119.

Saiz-Jimenez, C. and de Leeuw, J. W. 1986b. Lignin pyrolysis products: their structures and their significance as biomarkers. Organic Geochemistry 10:869-876.

Salzman, P. C. 1980. When nomads settle. Praeger, New York. 184 pp.

Sanchez, P. A. 1976. Properties and management of soils in the tropics. Wiley, New York. 618 pp.

Sanchez, P. A. and Benites, J. R. 1987. Low-input cropping for acid soils of the humid tropics. Science 238:1521-1527.

Sanchez, P. A. and Buol, S. W. 1975. Soils of the tropics and the world food crisis. Science 188:598-603.

Sanchez, P. and Miller, R. H. 1986. Organic matter and soil fertility management of acid soils in the tropics. Transactions of the 13th International Congress of Soil Science (Hamburg) 6:609-625.

Sanchez, P. and Salinas, J. G. 1981. Low-input technology for managing Oxisols and Ultisols in tropical America. Advances in Agronomy 34:279-406.

Sanchez, P., Gichuru, M. P., and Katz, L. B. 1982. Organic matter in major soils of the tropical and temperate regions. Transactions of the 12th International Congress of Soil Science (New Delhi) 1:99-114.

Sanchez, P., Villachica, J. H., and Bandy, D. E. 1983. Soil fertility dynamics after clearing a tropical rainforest in Peru. Soil Science Society of America Journal 47:1171-1178.

Sanford, R. L. Jr. 1985. Root ecology of mature and successional Amazon forests. Ph.D. dissertation. University of California, Berkeley.

Santantonio, D., Hermann, R. K., and Overton, W. S. 1977. Root biomass studies in forest ecosystems. Pedobiologia 17:1-31.

Sarmiento, G. 1984. The ecology of neotropical savannas. Harvard University Press, Cambridge, Massachusetts. 235 pp.

Sauerbeck, D. R. and Gonzalez, M. A. 1977. Field decomposition of carbon-14 labelled plant residues in various soils of the Federal Republic of Germany and Costa Rica. In: Soil organic matter studies. Proceedings of the FAO/IAEA Symposium, vol. 1. IAEA, Vienna. pp. 159-170.

Schaefer, D., Steinberger, Y., and Whitford, W. G. 1985. The failure of nitrogen and lignin control of decomposition in a North American desert. Oecologia (Berlin) 65:382-386.

Scharpenseel, H. W. and Schiffman, H. 1977. Radiocarbon dating of soils, a review. Zeitschrift für Pflanzenesnaehrung und Bodenkunde 140: 159-174.

Schnitzer, M. 1978. Humic substances: chemistry and reactions. In: M. Schnitzer and S. U. Khan (eds.), Soil organic matter. Elsevier, New York. pp. 1-58.

———. 1977. Recent findings on the characterization of humic substances extracted from soils from widely differing climatic zones. In: Soil organic matter studies, vol. II. International Atomic Energy Agency, Vienna. pp. 117-132.

———. 1972. Chemical, spectroscopic, and thermal methods for the classification and characterization of "humic substances." In: Proceedings of the International Meeting on Humic Substances. Wageningen. pp. 293-310.

Schnitzer, M. and Chan, Y. K. 1986. Structural characteristics of a fungal melanin and a soil humic acid. Soil Science Society of America Journal 50:67-71.

Schnitzer, M. and Khan, S. U. 1978. Soil organic matter. Elsevier, New York.

Schuppli, P. A. and McKeague, J. A. 1984. Limitations of alkali-extractable organic fractions as bases of soil classification criteria. Canadian Journal of Soil Science 64:173-186.

Schwartz, D., Mariotti, A., Lanfranchi, R., and Gulliet, B. 1986. $^{13}C/^{12}C$ ratios of soil organic matter as indicators of vegetation changes in the Congo. Geoderma 39:97-103.

Schwertmann, U., Kodama, H., and Fischer, W. O. 1986. Mutual interactions between organics and iron oxides. Chapter 7, in: P. M. Huang and M. Schnitzer (eds.), Interactions of soil minerals with natural organics and microbes. Special Publication 17. Soil Science Society of America, Madison, Wisconsin.

Sequi, P., Aringhieri, R., and Pardini, G. 1980. Effect of peroxidation on soil electrochemical properties. Zeitschrift für Pflanzenernahrung und Bodenkunde 143:298-305.

Sharpley, A. N. and Smith, S. J. 1983. Distribution of phosphorus forms in virgin and cultivated soils and potential erosion losses. Soil Science Society of America Journal 47:581-586.

Shaw, C. and Pawluk, S. 1986. Faecal microbiology of Octolasion tyrtaeum, Aporrectodea turgida and Lumbricus terrestris and its relation to the carbon budgets of three artificial soils. Pedobiologia 29:377-389.

Shen, S. M., Pruden, G., and Jenkinson, D. S. 1984. Mineralization and immobilization of nitrogen in fumigated soil and the measurement of microbial biomass nitrogen. Soil Biology and Biochemistry 16:437-444.

Shoji, S., Ito, T., Nakamura, S., and Saigusa, M. 1987. Properties of humus of Andosols from New Zealand, Chile and Ecuador. Japanese Journal of Soil Science and Plant Nutrition 58:4873-4879.

Siem, N. T., Fridland, V. M., and Orlov, D. S. 1977. Composition and properties of the humic substances of the major soils of North Vietnam. Pochvovedenie 8:39-54.

Singh, J. S. and Gupta, S. R. 1977. Plant decomposition and soil respiration in terrestrial ecosystems. Botanical Review 43:449-528.

Singh, J. S., Laurenroth, W. K., Hunt, H. W., and Swift, D. M. 1984. Bias and random errors in estimates of root production: a simulation approach. Ecology 65:1760-1764.

Skjemstad, J. O., Dalal, R. C., and Barron, P. F. 1986. Spectroscopic investigations of cultivation effects on organic matter of Vertisols. Soil Science Society of America Journal 50:354-359.

Smith, M. S., Frye, W. W., and Varco, J. J. 1987. Legume winter cover crops. Advances in Soil Science 7:95-139.

Smith, S. J. and Young, L. J. 1975. Distribution of nitrogen forms in virgin and cultivated soils. Soil Science 120:354-360.

Sobrado, M. A. and Medina, E. 1980. General morphology, anatomical structure and nutrient content of sclerophyllus leaves of the "Bana" vegetation of Amazonas. Oecologia (Berlin) 45:341-345.

Soil Survey Staff. 1987. Keys to Soil Taxonomy. SMSS Technical Monograph 6. Cornell University, Ithaca, New York.

————. 1975. Soil Taxonomy: a basic system of soil classification for making and interpreting soil surveys. USDA Department of Agriculture Handbook 436. Government Printing Office, Washington, D.C.

Sollins, P., Robertson, G. P., and Uehara, G. 1989. Nutrient mobility in variable- and permanent-charge soils. Biogeochemistry. In Press.

Sollins, P., Spycher, G., and Glassman, C. A. 1984. Net nitrogen mineralization from light- and heavy-fraction forest soil organic matter. Soil Biology and Biochemistry 16:31-38.

Sørensen, L. H. 1981. Carbon-nitrogen relationships during the humification of cellulose in soils containing different amounts of clay. Soil Biology and Biochemistry 13:313-321.

————. 1974. Rate of decomposition of organic matter in soil as influenced by repeated air drying-rewetting and repeated additions of organic matter. Soil Biology and Biochemistry 6:287-292.

Sowden, F. J., Chen, Y., and Schnitzer, M. 1977. The nitrogen distribution in soils formed under widely differing climatic conditions. Geochim. Cosmochim. Acta 41:1524-1526.

Sowden, F. J., Griffith, S. M., and Schnitzer, M. 1976. The distribution of nitrogen in some highly organic tropical volcanic soils. Soil Biology and Biochemistry 8:55-60.

Spain, A. V. and Hutson, B. R. 1983. Dynamics and fauna of the litter layers. In: Soils: an Australian viewpoint. CSIRO, Melbourne/Academic Press, London. pp. 611-623.

Spain, A. V., Isbell, R. F., and Probert, M. E. 1983. Soil organic matter. In: Soils: an Australian viewpoint. CSIRO, Melbourne/Academic Press, London. pp. 551-563.

Sparling, G. P. 1981. Microcalorimetry and other methods to assess biomass and activity in soil. Soil Biology and Biochemistry 13:93-98.

Sparling, G. P. and West, A. W. 1988a. A direct extraction method to estimate soil microbial C: calibration *in situ* using microbial respiration and ^{14}C-labelled cells. Soil Biology and Biochemistry 20:337-343.

Sparling, G. P. and West, A. W. 1988b. Modifications to the fumigation-extraction technique to permit simultaneous extraction and estimation of soil microbial C and N. Communications in Soil Science and Plant Analysis 19:327-344.

Sposito, G. 1984. The surface chemistry of soils. Oxford University Press, New York.

Sposito, G. and Schindler, P. W. 1986. Reactions at the soil colloid-soil solution interface. Transactions of the 13th International Congress of Soil Science (Hamburg) 6:683-699.

Spycher, G. and Young, J. L. 1977. Density fractionation of water-dispersible soil organic-mineral particles. Communications in Soil Science and Plant Analysis 8:37-48.

Spycher, G., Sollins, P., and Rose, S. L. 1983. Carbon and nitrogen in the light fraction of a forest soil: vertical distribution and seasonal patterns. Soil Science 135:79-87.

Stanford, G. and Epstein, E. 1974. Nitrogen mineralization-water relations in soils. Soil Science Society of America Proceedings 38:103-107.

Stanford, G. and Smith, S. J. 1972. Nitrogen mineralization potentials of soils. Soil Science Society of America Proceedings 36:465-472.

Stanford, G., Frere, M. H., and Schwaniger, D. E. 1973. Temperature coefficients of soil nitrogen mineralization. Soil Science 115:321-323.

Stefanson, R. C. 1971. Effect of periodate and pyrophosphate on the seasonal changes in aggregate stabilization. Soil Research 9:33-41.

Stevenson, F. J. 1986. Cycles of soil: carbon, nitrogen, phosphorus, sulfur, micronutrients. Wiley, New York. 380 pp.

———. 1982a. Humus chemistry: genesis, composition, reactions. Wiley, New York. 443 pp.

———. 1982b. Organic forms of soil nitrogen. In: F. J. Stevenson (ed.), Nitrogen in agricultural soils. American Society of Agronomy, Madison, Wisconsin. pp. 67-122.

Stevenson, F. J. and Fitch, A. 1986. Chemistry of complexation of metal ions with soil solution organics. In: P. M. Huang and M. Schnitzer. (eds.), Interactions of soil minerals with natural organics and microbes. Soil Science Society of America Special Publication 17. Madison, Wisconsin. pp. 29-58.

Stoop, W. 1980. Ion adsorption mechanisms in oxidic soils; implications for point of zero charge determinations. Geoderma 23:303-314.

Stott, D. E., Kassim, G., Jarrell, W. M., Martin, J. P., and Haider, K. 1983. Stabilization and incorporation into biomass of specific plant carbons during biodegradation in soil. Plant and Soil:15-26.

Stout, J. D. and Lee, K. E. 1980. Ecology of soil micro- and macro-organisms. In: B. K. G. Theng (ed.), Soils with variable charge. New Zealand Society of Soil Science, Lower Hutt. pp. 353-372.

Swift, M. J. (ed.) 1986. Tropical soil biology and fertility: inter-regional research planning workshop. International Union of Biological Sciences, Paris.

———. 1985. Tropical soil biology and fertility (TSBF). Planning for research. Special Issue 9. Biology International, International Union of Biological Sciences, Paris. 24 pp.

Swift, M. J. and Sanchez, P. A. 1984. Biological management of tropical soil fertility for sustained productivity. Nature and Resources 10:1-8.

Swift, M, J., Russel-Smith, A., and Perfect, T. J. 1981. Decomposition and mineral nutrient dynamics of plant litter in a regenerating bush fallow in the sub-humid tropics. Journal of Ecology 69:981-995.

Swift, M.J., Heal, O. W., and Anderson, J. M. 1979. Decomposition in terrestrial ecosystems. Blackwell Scientific, Oxford, England.

Swift, R. S. 1985. Fractionation of soil humic substances. In: G. R. Aiken, D. M. McKnight, R. L. Wershaw, and P. MacCarthy (eds.), Humic substances in soil, sediment, and water. Wiley, New York. pp. 397-408.

Szott, L. T. 1987. Improving the productivity of shifting cultivation in the Amazon Basin of Peru through the use of leguminous vegetation. Ph.D. dissertation. North Carolina State University, Raleigh. 168 pp.

Tabatabai, M. A. and Al-Khafaji, A. A. 1980. Comparison of nitrogen and sulfur mineralization in soils. Soil Science Society of America Journal 44:1000-1006.

Talpaz, H.P., Fine, P., and Bar-Yosef, B. 1981. On the estimation of N-mineralization parameters from incubation experiments. Soil Science Society of America Journal 45:993-996.

Tanner, E. V. J. 1981. The decomposition of leaf litter in Jamaican rain forests. Journal of Ecology 69:263-275.

Tate, K. R. 1984. The biological transformation of P in soil. Plant and Soil 76:245-256.

Tate, K. R. and Churchman, G. J. 1978. Organo-mineral fractions of a climosequence of soils in New Zealand tussock grasslands. Journal of Soil Science 29:331-339.

Tate, K. R. and Newman, R. H. 1982. Phosphorus fractions of a climosequence of soils in New Zealand tussock grassland. Soil Biology and Biochemistry 14:191-196.

Tate, K. R. and Theng, B. K. G. 1980. Organic matter and its interactions with inorganic soil constituents. Chapter 12, in: B. K. G. Theng (ed.), Soils with variable charge. New Zealand Society of Soil Science, Lower Hutt. pp. 225-249.

Tate, K. R., Ross, D. J., and Feltham, C. W. 1988. A direct extraction method to estimate soil microbial C: effects of experimental variables and some different calibration procedures. Soil Biology and Biochemistry 20:319-335.

Tate, K. R., Yamamoto, K., Churchman, G. J., Meinhold, R., and Newman, R.H. 1990. Relationships between the type and carbon chemistry of humic acids from some New Zealand and Japanese soils. Soil Science and Plant Nutrition. In Press.

Tate, R. L. III. 1987. Soil organic matter: biological and ecological effects. Wiley, New York. 291 pp.

Taylor, H. M. (ed.) 1987. Minirhizotron observation tubes: methods and applications for measuring rhizosphere dynamics. Special Publication 150. American Society of Agronomy, Madison, Wisconsin. 143 pp.

Tenney, G. and Waksman, S. A. 1929. Composition of natural organic materials and their decomposition. IV. The nature and rapidity of decomposition of the various organic complexes in different plant materials, under areobic conditions. Soil Science 28:55-84.

Tessens, E. and Zauyah, S. 1982. Positive permanent charge in Oxisols. Soil Science Society of America Journal 46:1103-1106.

Theng, B. K. G. 1987. Clay-humic interactions and soil aggregate stability. In: P. Rengasamy (ed.), Soil structure and aggregate stability. Seminar Proceedings. Institute of Irrigation and Salinity Research, Tatura, Australia. pp. 32-73.

———. 1979. Formation and properties of clay-polymer complexes. Developments in Soil Science 9. Elsevier, Amsterdam. 362 pp.

Theng, B. K. G., Churchman, G. J., and Newman, R. H. 1986. The occurrence of interlayer clay-organic complexes in two New Zealand soils. Soil Science 142:262-266.

Tiessen, H. and Stewart, J. W. B. 1988. Light and electron microscopy of stained microaggregates: the role of organic matter and microbes in soil aggregation. Biogeochemistry 5:312-322.

Tiessen, H. and Stewart, J. W. B. 1983. Particle size fractions and their use in studies of soil organic matter. II. Cultivation effects on organic matter composition in size fractions. Soil Science Society of America Journal 47:509-514.

Tiessen, H., Karamanos, R. H., Stewart, J. W. B., and Selles. F. 1984a. Natural nitrogen-15 abundance as a indicator of soil organic matter transformations in native and cultivated soils. Soil Science Society of America Journal 48:312-315.

Tiessen, H., Stewart, J. W. B., and Cole, C. V. 1984b. Pathways of phosphorus transformations in soils of differing pedogenesis. Soil Science Society of America Journal 48:853-858.

Tiessen, H., Stewart, J. W. B., and Hunt, H. W. 1984c. Concepts of soil organic matter transformations in relation to organo-mineral particle size fractions. Plant and Soil 76:287-295.

Tiessen, H., Stewart, J. W. B., and Moir, J. O. 1983. Changes in organic and inorganic phosphorus composition of two grassland soils and their particle size fractions during 60-90 years of cultivation. Journal of Soil Science 34:815-823.

Tinker, P. B. H. and Ziboh, C. O. 1959. A study of some typical soils supporting oil palms in southern Nigeria. Journal of the West African Institute for Oil Palm Research 3:16-51.

Tipping E. 1981. Adsorption by goethite (α-FeOOH) of humic substances from three different lakes. Chemical Geology 33:81-89.

Tisdale, S. L. and Nelson, W. E. 1956. Soil fertility and fertilizers. Macmillan, New York.

Tisdall, J. M. and Oades, J. M. 1982. Organic matter and water-stable aggregates in soils. Journal of Soil Science 33:141-163.

Tisdall, J. M. and Oades, J. M. 1980. The effect of crop rotation on aggregation in a red-brown earth. Australian Journal of Soil Research 18:423-433.

Tiwari, K. N., Tiwari, S. P., and Pathak, A. N. 1980. Studies on green manuring of rice in double cropping system in a partially reclaimed saline sodic soil. Indian Journal of Agronomy 25:136-145.

Toreu, B. N., Thomas, F. G., and Gillman, G. P. 1988. Phosphate sorption characteristics of soils of the north Queensland coastal region. Australian Journal of Soil Research 26:465-477.

Toutain, F. 1987. Les litières: siège de systèmes interactifs et moteur de ces interactions. Revue d'Ecologie et de Biologie du Sol 24:231-242.

Trapnell, C. G., Friend, M. T., Chamberlain, G. T., and Birch, H. F. 1976. The effects of fire and termites on a Zambian woodland soil. Journal of Ecology 64:577-588.

TropSoils. 1987. Technical report for 1985-1986. North Carolina State University, Raleigh.

Tsutsuki, K., Suzuki, C., Kuwatsuka, S., Becker-Heidmann, P., and Scharpenseel, H. W. 1988. Investigation on the stabilization of the humus in Mollisols. Zeitschrift für Pflanzenernahrung und Bodenkunde 151:87-90.

Turchenek, L. W. and Oades, J. M. 1979. Fractionation of organio-mineral complexes by sedimentation and density techniques. Geoderma 21:311-343.

Tweneboah, C. K., Greenland, D. J., and Oades, J. M. 1967. Changes in charge characteristics of soils after treatment with 0.5 M calcium chloride pH 1.5. Australian Journal of Soil Research 5:247-261.

Uehara, G. and Gillman, G. 1981. The mineralogy, chemistry and physics of tropical soils with variable charge clays. Westview, Boulder, Colorado. 170 pp.

Vallis, I. and Jones, R. J. 1973. Net mineralization of nitrogen in leaves and leaf litter of *Desmodium intortum* and *Phaseolus atropurpureus* mixed with soil. Soil Biology and Biochemistry 5:391-398.

van Veen, J. A. and Paul, E. A. 1981. Organic C dynamics in grassland soils. I. Background information and computer simulation. Canadian Journal of Soil Science 61:185-201.

van Veen, J. A., Ladd, J. N., and Frissel, M. J. 1984. Modeling C and N turnover through the microbial biomass in soil. Plant and Soil 76:257-274.

Verstraete, W. and Voets, J. P. 1976. Nitrogen mineralization tests and potentials in relation to soil management. Pedologie 26:15-26.

Vitousek, P. M. 1984. Litterfall, nutrient cycling, and nutrient limitation in tropical forests. Ecology 65:285-298.

Vitousek, P. M. and Matson, P. A. 1988. Nitrogen transformations in a range of tropical forest soils. Soil Biology and Biochemistry 20:361-367.

Vitousek, P. M. and Sanford, R. L. 1986. Nutrient cycling in moist tropical forests. Annual Review of Ecology and Systematics 17:137-167.

Vitousek, P. M., Matson, P. A., and Turner, D. R. 1989. Elevational and age gradients in Hawaiian montane rainforest: foliar and soil nutrients. Tropical Mountain Ecosystem Program of the IUBS Decade of the Tropics. In Press.

Vogt, K. A., Grier, C. C., and Vogt, D. J. 1986. Production, turnover, and nutrient dynamics of above- and below-ground detritus of world forests. Advances in Ecological Research 15:303-377.

Vogt, K. A., Edmonds, R. L., Grier, C. C., and Piper, S. R. 1980. Seasonal changes in mycorrhizal and fibrous-textured root biomass in 23- and 180-year-old Pacific silver fir stands in western Washington. Canadian Journal of Forest Research 10:523-529.

Vogt, K. A., Grier, C. C., Meier, C. E., and Edwards, R. L. 1982. Mycorrhizal role in net primary production and nutrient cycling in *Abies amabilis* (Dougl). Forbes ecosystems in western Washington. Ecology 63:370-380.

Volkoff, B. and Cerri, C. C. 1987. Carbon isotope fractionation in subtropical Brazilian grassland soils. Comparison with tropical forest soils. Plant and Soil 102:27-31.

Wada, K. (ed.) 1986. Ando soils in Japan. Kyushu University Press, Fukuoka.

———. 1985. The distinctive properties of Andosols. In: B. A. Stewart (ed.), Advances in soil science, vol. 2. Springer-Verlag, New York.

―――. 1980. Mineralogical characteristics of Andisols. In: B. K. G. Theng (ed.), Soils with variable charge. New Zealand Society of Soil Science, Lower Hutt. pp. 87-107.

Wada, K. and Aomine, S. 1973. Soil development on volcanic materials during the Quaternary. Soil Science 116:170-177.

Wada, K. and Higashi, T. 1976. The categories of aluminum and iron humus complexes in ando soils determined by selective dissolution. Journal of Soil Science 27:357-368.

Wade, M. K. and Sanchez, P. A. 1983. Mulching and green manure applications for continuous crop production in the Amazon Basin. Agronomy Journal 75:39-45.

Wagner, G. H. and Mutakar, V. V. 1968. Amino components of soil organic matter formed during humification of ^{14}C-glucose. Soil Science Society of America Proceedings 32:683-686.

Waid, J. S. 1975. Hydroxamic acids in soil systems. Chapter 3, in: E. A. Paul and A. D. McLaren (eds.), Soil biochemistry, vol. 4. Dekker, New York.

Waksman, S. A. 1936. Humus. Williams and Wilkins, Baltimore.

Werger, M. J. A. and Ellenbroek, G. A. 1978. Leaf size and leaf constituence of a riverine forest formation along a climatic gradient. Oecologia (Berlin) 34:297-308.

Wershaw, R. L. 1985. Application of nuclear magnetic resonance spectroscopy for determining functionality in humic substances. In: G. R. Aiken, D. M. McKnight, R. L. Wershaw, and P. MacCarthy (eds.), Humic substances in soil, sediment, and water. Wiley, New York. pp. 561-582.

Wetselaar, R. and Ganry, F. 1982. Nitrogen balance in tropical agroecosystems. In: Y. R. Dommergues and H. G. Diem (eds.), Microbiology of tropical soils and plant productivity. Martinus Nijhoff, Dordrecht, The Netherlands. pp. 1-36.

White, R. E. 1988. Leaching. In: J. R. Wilson (ed.), Advances in nitrogen cycling in agricultural ecosystems. CAB International, Wallingford, U.K. pp. 193-210.

Wielemaker, W. G. 1984. Soil formation by termites: a study in the Kisii area, Kenya. Ph.D. thesis. Agricultural University, Wageningen, The Netherlands.

Wilding, L. P. and Dees, L. R. 1978. Spatial variability: a pedologist's viewpoint. American Society of Agronomy Special Publication 34:1-12. American Society of Agronomy, Madison, Wisconsin.

Wilson, J. R. (ed.) 1988. Advances in nitrogen cycling in agricultural ecosystems. CAB International, Wallingford, U.K.

Wilson, M. A. 1987. NMR techniques and applications in geochemistry and soil chemistry. Pergamon, Oxford. 353 pp.

———. 1981. Application of n.m.r. spectroscopy to the study of soil organic matter. Journal of Soil Science 32:167-186.

Wilson, M. A., Barron, P. F., and Goh, K. M. 1981a. Cross polarisation ^{13}C-N.M.R. spectroscopy of some genetically related New Zealand soils. Journal of Soil Science 32:419-425.

Wilson, M. A., Barron, P. F., and Goh, K. M. 1981b. Differences in structure of organic matter in two soils as demonstrated by ^{13}C cross-polarisation nuclear magnetic resonance spectroscopy with magic angle spinning. Geoderma 26:323-327.

Wilson, M. A., Heng, S., Goh, K. M., Pugmire, R. J., and Grant, D. M. 1983. Studies of litter and acid-insoluble soil organic matter fractions using ^{13}C cross polarization nuclear magnetic resonance spectroscopy with magic angle spinning. Journal of Soil Science 34: 83-97.

Wilson, M. A., Pugmire, R. J., Zilm, K. W., Goh, K. M., Heng, S., and Grant, D. M. 1981c. Cross-polarization ^{13}C-NMR spectroscopy with 'magic-angle' spinning characterizes organic matter in whole soils. Nature 294:648-650.

Wolt, J. and Graveel, J. G. 1986. A rapid routine method for obtaining soil solution using vacuum displacement. Soil Science Society of America Journal 50:602-605.

Wood, T. G. 1988. Termites and the soil environment. Biology and Fertility of Soils 6:228-236.

Wood, T. G., Johnson, R. A., and Anderson, J. M. 1983. Modification of soils in Nigerian savanna by soil-feeding Cubitermes (Isoptera, Termitidae). Soil Biology and Biochemistry 15:575-579.

Yamamoto, K., Tate K. R., and Churchman, G. J. 1989. A comparison of the humus substances from some volcanic ash soils in New Zealand and Japan. Soil Science and Plant Nutrition 35. In Press.

Yamamoto, K., Tate, K. R., and Churchman, G. J. 1987. A comparison of the humus in, and humic acids from, some New Zealand and Japanese soils. Laboratory Report BB4. New Zealand Soil Bureau, Lower Hutt. 22 pp.

Yasuhara, K. and Takenaka, H. 1977. Physical and mechanical properties. In: T. Yamanouchi (ed.), Engineering problems of organic soils in Japan. JSSMFE, Tokyo.

Young, A. 1976. Tropical soils and soil survey. Cambridge University Press, Cambridge, England. 322 pp.

Zech, W., Johansson, M.-J., Haumaier, L., and Malcolm, R. L. 1987. CPMAS ^{13}C NMR and IR spectra of spruce and pine litter and of the Klason lignin fraction at different stages of decomposition. Zeitschrift für Pflanzenernahrung und Bodenkunde 150:262-265.

Zech, W., Koegel, I., Zucker, A., and Alt, H. 1985. CP-MAS-^{13}C-NMR-Spektren organischer Lagen einer Tangelrendzina. Zeitschrift für Pflanzenernahrung und Bodenkunde 148:481-488.

Zeikus, J. G. 1981. Lignin metabolism and the carbon cycle: polymer biosynthesis, biodegradation, and environmental recalcitrance. In: M. Alexander (ed.), Advances in microbial ecology. Plenum Press, New York and London. pp. 211-243.

Zucker, W. V. 1983. Tannins: does structure determine function? An ecological perspective. American Naturalist 121: 335-365.

Zunino, H., Borie, F., Aguilera, S., Martin, J. P., and Haider, K. 1982. Decomposition of ^{14}C-labelled glucose, plant and microbial products and phenols in volcanic ash derived soils of Chile. Soil Biology and Biochemistry 14: 37-43.

Full References for Cited
Tables and Figures

Chapter: 1

Figure: 1 - A.V. Spain, R.F. Isbell, and M.E. Probect. 1983. Soil organic matter. In: Soils: An Australian viewpoint. CSIRO, Melbourne/Academic Press, London. pp. 551-563.

Figures: 2 & 4 - J.M. Anderson and M.J. Swift. 1983. Decomposition in tropical forests. In: S.L. Sutton, T.C. Whitmore, and A.C. Chadwick (eds.), Tropical Rainforest: Ecology and Management. Special Publication 2, British Ecology Society. Blackwell Scientific, Oxford, England. pp. 287-309.

Figures: 3 & 5 - S. Brown and A.E. Lugo. 1982. The storage and production of organic matter in tropical forests and their role in the global carbon cycle. Biotropica 14:161-187.

Figure: 6 - H. Peterson. 1982b. The Total Soil Fauna Biomass and Its Composition. Oikos, Volume 39, Number 3, pp. 330-339. Munksgaard/Copenhagen/1987.

Figure: 9 - S. Shoji, T. Ito, S. Nakamura, and M. Saigusa. Properties of Humus of Andisols from New Zealand, Chile and Ecuador. Japanese Journal of Soil Science and Plant Nutrition, Vol. 58, pp. 473-479. Japan Society of Soil Science and Plant Nutrition/Tokyo/1987.

Figure: 11 - J.M. Oades, A.M. Vasallo, A.G. Waters, and M.A. Wilson. 1987. Characterization of organic matter in particle size and density fractions from a red-brown earth by solid-state ^{13}C NMR. Australian Journal of Soil Research 25:71-82.

Figure: 12 - B.K.G. Theng, G.J. Churchman, and R.H. Newman. 1986. The Occurrence of Interlayer Clay-Organic Complexes in Two New Zealand Soils. Soil Science 142:262-266.© by Williams and Wilkins, 1986.

Figure: 14 - J. Balesdent, G.H. Wagner, and A. Mariotti. 1988. Soil organic matter turnover in long-term field experiments as revealed by carbon-13 natural abundance. Soil Science Society of America Journal 52:118-124.

Chapter: 2

Figure: 1 - W.B. Gonzalez and D.K. Sauerbeck. 1982. Decomposition of ^{14}C labelled plant residues in different soils and climates of Costa Rica. Proceedings of the Regional Colloquium on Soil Organic Matter Studies, Piracicaba, SP-Brazil. PROMOCET, Sao Paulo. Available through Centro Energia Nuclear na Agricultura, Caixa Postal 96, 13400 Piracicaba, SP, Brazil. pp. 141-146.

Figure: 2 - W.B. McGill and R.J.K. Myers. 1987. Controls on dynamics of soil and fertilizer nitrogen. pp. 73-99. In: R.F. Follett, J.W.B. Stewart, and C.V. Cole (eds.), Soil Fertility and Organic Matter as Critical Components of Production Systems. SSSA Special Publication 19. Soil Science Society of America, Inc.

Table: 2 - E.A. Paul and R.P. Voroney. 1983. Field interpretation of microbial biomass activity measurements. In: M.J. Klug and C.A. Reddy (eds.), Current Perspectives in Microbial Ecology. Michigan State University, American Society for Microbiology.

Table: 4 - Doran, J. W. 1987. Tillage effects on microbiological release of soil organic nitrogen. In: T. J. Gerikand and B. L. Harris (eds.), Conservation Tillage. Southern Regional No-Till Conference, July 1987. Texas A& M University, College Station. pp. 63-66.

Chapter: 5

Figure: 2 - A.I. Khatibu, R. Lal, and R.K. Jara. 1984. Effects of tillage methods and mulching on erosion and physical properties of a sandy clay loam in an Equatorial warm humid region. Field Crops Research 8:239-254. Elsevier Science Publishers. Physical Sciences and Engineering Division.

Figures: 3 & 4 - T.L. Lawson and R. Lal. 1979. Response of maize to surface and buried straw mulch on a tropical Alfisol. In: R. Lal (ed.), Soil Tillage and Crop Production. Proceedings Series 2. IITA, Ibadan, Nigeria. pp. 63-74.

Chapter: 6

Tables: 1 & 2 - W.J. Parton, D.S. Schimel, C.V. Cole, and D.S. Ojima. 1987. Analysis of Factors Controlling Soil Organic Matter Levels in Great Plains Grasslands. Volume 51, pp. 1173-1179. Soil Science Society of America Journal.

Figure: 3 - W.J. Parton, J.W.B. Stewart, and C.V. Cole. 1988. Dynamics of C, N, P, and S in grassland soils: a model. Biogeochemistry 5:109-131. Martinus Nijhoff Publishers.

Index

A

Adsorption of
  humic substances, 78
  organic acids, 76
  organic cations, 77
  phosphate, 90
  sulfate, 91
Aggregate size, 39
Aggregation of soil
  influence on mineralization
    of OM, 186
  mechanisms, 184
  polysaccharides and, 181, 185
Agricultural systems
  cropping systems, 121
Al/Fe complexes, 14, 20, 28
Allophane, 70, 84
Andosols
  See Soil Orders
Anion exchange capacity (AEC), 73
Aromaticity of humic acids, 189

B

Bacteria, 106
Biomass
  fauna, 12-13
  microbial, 14 - 15, 44
  of phosphorus, 191
  plant, 16

C

Carbohydrates, 9
Carbon
  alkyl, 12, 24, 26 - 27
  aromatic, 12, 20, 24
  available, 14, 17
  C/N ratio, 155, 160 - 169
  C/P ratio, 160 - 169
  carboxyl, 12, 22 - 23
  comparisons, 155
  cycling, 157
  effects of elevation on levels of, 162
  O-alkyl-, 24, 26
  to nitrogen ratio, 11, 14, 17, 25
Carbon-13
  nuclear magnetic resonance spectroscopy,
    186
Carbon-14, 185
Carbon-nitrogen ratio, 191-192
Carbon-nitrogen-phosphorus-sulfur
    ratio, 191
Carbon-phosphorus ratio, 191-192
Carbon-sulfur ratio, 191
Cation exchange capability (CEC), 73, 88
Cation exchange capacity, 62
CENTURY
  Hawaii forest sites, 162
  major input variable required for, 161
  nutrient submodels, 161
  parameterization procedure, 162

P

R

Dynamics of SOM in Tropical Ecosystem

S

Savanna
    resource use by earthworms, 116
Soil amendments
    organic inputs as, 147
Soil invertebrates
    direct effects, 112
    indirect effects, 112
Soil orders
    Andisols, 19, 28, 30, 70,75, 84
    Andosols, 70, 75, 91
    Mollisols, 27
    Oxisols, 19, 23, 70 75, 84, 88, 90, 91-92
    Spodosols, 75, 84
    Ultisols, 6, 23, 75
    Vertisols, 19, 23
Soil pH, 90
SOM fractions
    carbon input, 117
SOM mineralization, 117
Stable carbon isotopes, 28

T

Temperate soils, 6, 17
Temporal & spatial distribution
    of root litter, 141
    See also Root litter

Termites and biogeography, 113
    effects on soils, 113
    effects on SOM, 114
    feeding ecology, 113
    litter consumption, 114
Tracer studies
    of carbon, 185, 194
    of nitrogen, 185, 194
Tropical forests, 10, 19
Tropical soils, 6, 14, 17, 28, 30
Tropical SOM model, 153

U

Ultisols
    See Soil Orders

V

Variable charge, 71
Vertisols
    See Soil Orders

W

Waterlogged, 105, 106